OPERATIONAL RISK MANAGEMENT

THE INTEGRATION OF DECISION, COMMUNICATIONS, AND MULTIMEDIA TECHNOLOGIES

GIAMPIERO E.G. BEROGGI

WILLIAM A. WALLACE

OPERATIONAL RISK MANAGEMENT

THE INTEGRATION OF DECISION, COMMUNICATIONS, AND MULTIMEDIA TECHNOLOGIES

by

GIAMPIERO E.G. BEROGGI

Delft University of Technology

Delft, The Netherlands

and

WILLIAM A. WALLACE

Rensselaer Polytechnic Institute,

Troy, New York, U.S.A.

Kluwer Academic Publishers

Boston/Dordrecht/London

Distributors for North, Central and South America:
Kluwer Academic Publishers
101 Philip Drive
Assinippi Park
Norwell, Massachusetts 02061 USA

Distributors for all other countries:
Kluwer Academic Publishers
Distribution Centre
Post Office Box 322
3300 AH Dordrecht, THE NETHERLANDS

Library of Congress Cataloging-in-Publication Data

A C.I.P. Catalogue record for this book is available
from the Library of Congress.

Copyright © 1998 by Kluwer Academic Publishers

All rights reserved. No part of this publication may be reproduced, stored in a retrieval system or transmitted in any form or by any means, mechanical, photo-copying, recording, or otherwise, without the prior written permission of the publisher, Kluwer Academic Publishers, 101 Philip Drive, Assinippi Park, Norwell, Massachusetts 02061

Printed on acid-free paper.

Printed in the United States of America

To Jean and Penny

TABLE OF CONTENTS

Preface   xi

Acknowledgments   xv

CHAPTER I: THE GENESIS OF OPERATIONAL RISK MANAGEMENT

1. Closing the Gap   1
   1.1 Introduction   1
   1.2 Motivation for Operational Risk Management   3
   1.3 Operational Control: Definition, Tasks, and State of the Art   8

2. Advanced Communications Technologies   12
   2.1 Introduction   12
   2.2 Overview of Satellite Communications Technology   15
   2.3 Applications of Communications Technologies   18
   2.4 Outlook   21

3. The Concept of Operational Risk Management   24
   3.1 Basic Concept   24
   3.2 Practicality of Operational Risk Management for Transportation Systems   31

CHAPTER II: MATHEMATICAL EXPOSITION OF OPERATIONAL RISK MANAGEMENT

1. The Graph Theoretic Approach of ORM   38
   1.1 The Topological Graph Structure   38
   1.2 Solvability of Operations   41
   1.3 Preference Classes   43

2. Individual Decision Maker Situation   47
   2.1 Cognitive Assumptions for Preference Assessment   47
   2.2 Preference Algebra and Reasoning Logic   51
   2.3 Finding and Revising Optimal Courses of Action   53
   2.4 Relation to Multiattribute Utility Model   57
   2.5 The Process of Preference Assessment and Decision Making   58

3. Multiple Expert Decision Making Situation   60
   3.1 The Decision Process for Multiple Experts   60
   3.2 An Adaptive Model for Cardinal Preference Assessments   62
   3.3 An Adaptive Model for Ordinal Preference Assessments   66
   3.4 Choice Process   68

CHAPTER III: ADVANCES IN INFORMATION TECHNOLOGIES

1. Multimedia and Hypermedia 71
  1.1 Multimedia 71
    A) Multimedia on the Network 71
    B) QuickTime 72
  1.2 The Hypermedia Environment 73
    A) Basic Concept 73
    B) The Programming Environment 75
    C) Systems Architecture and Functionality 77

2. Internet 79
  2.1 Introduction 79
  2.2 The Architecture of an Extended MEMIS 80
  2.3 The MEMIS Modules 82
  2.4 Tasks to be Addressed in MEMIS 87
  2.5 Conclusions 88

3. Virtual Reality 89
  3.1 Introduction 89
  3.2 Components of Virtual Reality 91
    A) Taxonomy 91
    B) Virtual Worlds and Cyberspace 92
    C) Technology in Virtual Reality 93
    D) The Role of Information Technology 94
  3.3 Impact on Emergency Management 95
    A) From CAMEO to Virtual World Navigation 95
    B) Tele-Virtual Conferencing in Cyberspace 96
    C) Training in Cyberspace 97
    D) Teleoperation and Telepresence 98
  3.4 Conclusions 99

CHAPTER IV: ROUTING OF HAZARDOUS MATERIALS

1. The ORM Environment 100
  1.1 Introduction 100
  1.2 Operational Risk Management for
    Transportation of Hazardous Materials 101

2. ORM Decision Logic for Transportation of Hazardous Materials 107
  2.1 General Aspects 107
  2.2 Ordinal Preference Routing Model 108
  2.3 Multiattribute Utility Routing Model 108
  2.4 Decision Making 109
  2.5 Example of Preference Routing Model 111

3. Decision Support Systems for Transportation of Hazardous Materials 112
   3.1 Capital District of New York State, USA 112
      A) Components of the Decision Support System 112
      B) Operational Risk Assessment for Transportation of Hazardous Materials 115
      C) Route Selection Process 118
      D) Assessment of the DSS 119
   3.2 Switzerland 120
      A) Components of the Decision Support System 120
      B) Scenario Generation: Vehicles and Events 121
      C) Robustness of Routes 122
   3.3 Conclusions 123

CHAPTER V: EMERGENCY RESPONSE

1. The ORM Environment 124
   1.1 Introduction 124
   1.2 Operational Risk Management in Emergency Management 125

2. ORM Decision Logic for Emergency Response 128
   2.1 The Graph Model 128
   2.2 Values of Activities and Computing Optimal CAs 129
   2.3 Revising Courses of Actions 132
   2.4 An Illustrative Example 133

3. Decision Support System 136
   3.1 The Modeling Environment 136
   3.3 Preference Assessment and Decision Making 137
   3.4 Conclusions 139

CHAPTER VI: OPERATIONAL CONTROL

1. Air Raid Command 140
   1.1 Introduction 140
   1.2 The ORM Environment 140
   1.3 ORM Decision Logic for Air Raid Command 142
   1.4 Decision Support System 144
      A) The Concept 144
      B) Contingency Tasks 146
      C) Display 148
      D) An Illustrative Example 149
      E) Conclusions 150

2. Emergency Response at a Nuclear Power Generation Facility 151
   2.1 Introduction 151
   2.2 The ORM Environment 152

2.3 ORM Decision Logic for Emergency Response
    at Nuclear Power Plant ... 154
    A) Preference Structure ... 154
    B) Preference Ranking with Multiple Groups of Experts ... 155
2.4 Decision Support System ... 157
    A) Functions of the Prototype System ... 158
    B) Emergency Responses ... 159
    C) Assessment ... 161

3. Concluding Comments ... 162

CHAPTER VII: ASSESSMENT OF OPERATIONAL RISK MANAGEMENT

1. Experimental Assessment of the Decision Logic ... 163
  1.1 Introduction ... 163
  1.2 The Decision Problem and Four Decision Models ... 164
  1.3 Hypotheses ... 167
    A) Hypotheses Concerning Effort (HE) ... 168
    B) Hypotheses Concerning Accuracy (HA) ... 169
  1.4 The Experiment ... 170
    A) Data for the Task ... 170
    B) Three Scenarios ... 171
    C) Subjects ... 173
  1.5 Results ... 175
    A) Overview ... 175
    B) Results Concerning Effort ... 177
    C) Results Concerning Accuracy ... 179
    D) Effort and Accuracy ... 181
    E) Task Complexity and Accuracy ... 182
    F) Cognitive Load ... 183
    G) Final Judgment ... 185
    H) Additional Results ... 186

2. Simulation of Multi-Expert Decision Making ... 187
  2.1 Preference Aggregation ... 187
  2.2 Choices ... 190

3. Conclusions ... 191

References ... 196

Subject Index ... 206

Preface

Traditional approaches to risk management focus on strategic planning issues, such as the siting of hazardous material facilities, the designation of routes for hazardous material shipments, and the development of emergency response plans. The two major issues in risk management are: is the system safe enough, and which of the feasible options is the "best" one. Typical risk analysis methods, such as described in the many guidelines, require extensive human resources and several weeks of work. The results are used to design and operate new technological systems or improve existing facilities.

However, traditional risk management has focused on pre-event and post-event activities, such as regulations, training, design, strategic planning, risk assessment, and emergency response. Little attention has been paid to safety considerations during the hazardous operations. The potential for catastrophic loss inherent in hazardous operations calls for closing the gap between pre-event and post-event actions by controlling the hazardous operations in real-time in terms of operational risk management. Technological advances have demonstrated the feasibility for operational risk management and economical considerations its realism. A control center supported by expert system technology including voice generation and flat panel display is no longer merely a vivid imagination of audacious decision analysts.

Operational risk management is not an alternative to strategic planning. Rather, it is a complement for issues that cannot be addressed from a strategic planning perspective. It is therefore obvious that any analytic approach to operational decision making must be in accordance with the strategic considerations, and vice versa. Ideally, operational decision making should consist of following strategically specified courses of action. However, there will always be the need to address the unanticipated, resulting in the need for decision aids that support decision making required to respond to these events.

The need to control and guide remote and mobile operations from the more safe and secure headquarters is not new. However, limitations in information and communications technologies hindered managers in their ability to control these operations and respond to unexpected events. Only the implementation of satellite-based communications and positioning systems, and increased computer power have made real-time risk management possible.

Satellite tracking systems, coupled with communications, provide centralized headquarters with the ability to monitor and control remote and mobile operations in real-time. Satellite position systems can be used to determine the location of the mobile units and animated simulation can be employed to illustrate on-going operations.

Despite these remarkable innovations, the user of these systems often realized that there is a gap between the tasks that need to be addressed in an operational setting and the commercially available technology. Satellite tracking systems are built to provide a service that gives the user an economic advantage and are not designed for the safety and security of operations. Simulation environments provide high flexibility but little support for modeling safety and emergency response systems. Geographic information systems are multi-purpose but do not integrate their displays with local and regional emergency plans in a way that permits dynamic revisions and development of new courses of action. Expert systems have been used in support of emergency management tasks, but not incorporated into satellite systems. Finally, hyper- and multimedia environments provide very useful human-machine interfaces for decision support systems but there is still the need for the specification and development of the analytic procedures and models for risk managers.

To close the gap between the decision tasks that need to be addressed in an operational environment and the employment of these new technologies, appropriate decision support models must be developed. These models must aid the risk manager in both sensing the current state of the operations and reasoning about actions that need to be taken. Sensing is supported by advanced data acquisition, positioning, and communications technologies. Examples are mobile sensors in the field that gather data about the status of the operation, weather monitoring systems, satellite location systems to determine the coordinates of mobile units, satellite communications systems, and advanced visualization systems such as geographic information systems and multimedia technology.

Reasoning is supported by appropriate decision models. These models can include simulation models to study and display potential impacts based on dispersion plumes, expert system modules to infer possible consequences, data base systems to extract relevant facts about the operations, and symbolic and numerical models to compute new courses of action.

The issues to be addressed at the organizational level refer to investments in new technologies, changes of traditional work procedures, and definition of responsibilities for these new decision situations. Despite the many advantages that these new technologies seem to promise, there is a high degree of uncertainty about their economic and safety benefits. Many organizations involved in hazardous operations do not have the resources to take substantial financial risks. In addition, traditional work approaches are difficult to change because they have worked in the past. Any new approach must provide enough benefits to overcome this hindrance.

The outline of the book is as follows: Chapter I provides the raison d'être for the book, building on themes of advanced technologies in communications and computing and large-scale distributed enterprises which will increase the potential for catastrophic impacts on the enterprise and its environment.

Chapter II provides a mathematical exposition of the logic for operational risk management for a single decision maker. Then, the concept is extended to include multiple experts providing support to the individual decision maker. Finally, a structure for group decision making is provided.

Chapter III contains a review of the technologies that can provide the support needed to deal with these unexpected events, including hypermedia, multimedia, communications technologies, networking, compressed video, video conferencing, Internet, and virtual reality.

Chapters IV, V, and VI discuss applications of operational risk management in transportation of hazardous materials, emergency management, and operational control.

Chapter VII discusses an assessment of these new technologies by experimentation and simulation for individual and multiple experts. Conclusions are drawn with possible extensions to the logic of operational risk management in light of new technologies such as the Internet for access to knowledge.

The material used in this book is based on past work by the authors, most of which has been published previously in scholarly journals. The appropriate references of the authors' work for the seven chapters can be found in Part A) of the references. To avoid overload, no specific references to the authors' work are made in the text.

Giampiero E.G. Beroggi and William A, Wallace
Haarlem, The Netherlands and Troy, New York, U.S.A.

Acknowledgments

This book reports on research that we conducted over the last eight years at Rensselaer Polytechnic Institute (Troy, New York, USA), Swiss Federal Institute of Technology (Zürich, Switzerland), and Delft University of Technology (Delft, Netherlands).

Our special thanks go to people and institutions who provided financial and other support, including William Anderson at the U.S. National Science Foundation and Professor Willy A. Schmid at the Department of Local, Regional, and National Planning at ETH-Zürich. In addition, support from the Polyproject *Risk and Safety of Technological Systems* at ETH Zürich and the Dispatching School in Wil enabled us to conduct our field research in Switzerland. Our two home institutions, Rensselaer and TU Delft, provided an environment conducive to conducting both theoretical studies and implementation of the resulting logics in decision support technologies. We also wish to acknowledge and thank various scholarly journals for letting us publish in revised and expanded form most of the material in this book, especially Professor Andrew P. Sage, Editor of the *IEEE Transactions on Systems, Man, and Cybernetics*.

We are especially grateful to the collaborators who participated in different parts of the research projects: Markus Aebi, Anna M. Hersperger, Yasushi Ikeda, Laurie Waisel, Matthias Wiedmer, Martin B. Zumsteg, as well as a large number of colleagues, anonymous journal referees, and friends at their respective institutions. A warm thank goes to Jean Wallace and Penny Spring for their support, encouragement, and the good times we had in Cerentino, Haarlem, Nice, Paris, Troy, and Zürich.

CHAPTER I: THE GENESIS OF OPERATIONAL RISK MANAGEMENT

1. Closing the Gap

1.1 Introduction

During the last decades of the twentieth century, there has been an increasing concern with hazards and their associated risk to humans, property and the environment. There are at least three reasons for this phenomenon. First, these hazards are now being viewed as in some sense "manageable" because either their cause or their impact is due to human initiated activities. They are recognized as due to and perhaps part of a complex web of interactions among people, technologies and the environment, with multiple causes and consequences. Second, ambitious programs are in place and being planned by national and international organizations designed to provide the science needed to deal with these hazards. Finally, there is over fifty years of research on human response to hazards [Mitchell, 1990].

The focus of our research is on hazards that are due to sudden onset events with a very low probability of occurrence that have the potential for catastrophic impact on humans, property or the environment. Furthermore, these events occur during "operations," ongoing activities whose disruption could have serious consequences. Examples of technological operations with the potential for low probability, high consequence (**LP-HC**) events are transportation of hazardous materials, operations of chemical plants, and operation of nuclear power plants. The potential for LP-HC events is not only present during these routine operations, but also during non-routine activities such as maintenance, training, and emergency response operations.

The need for better ways to deal with the potential for catastrophic loss inherent in these types of operations has been widely recognized and accepted by both government and industry. However, there is justifiable debate over the means for improving safety in hazardous operations. Traditional approaches have focused on **pre-event activities** (e.g., systems design, regulations, training, risk assessment, and strategic planning) and **post-event activities** (e.g., emergency response and recovery). Pre-event activities are designed to reduce the risks before the operations occur, whereas post-event actions seek quick and efficient ways to minimize impacts when accidents and other disruptions occur.

The public and private organizations responsible for the hazardous operations recognize the limitations of these measures, both in theory and in practice.

The inability to have universal use of 'safe' operational procedures is due to the highly competitive nature of industry which limits, especially for smaller companies, the introduction of new technologies for improvements in safety. Regulations can be ineffectual due to the variety of public authorities involved and to problems with enforcement.

Training of the personnel is seen as one of the most promising ways to improve safety, since many LP-HC events are in some way caused by human error. However, even properly trained operators, such as response personnel, drivers, and pilots are all too often involved in situations which go beyond their cognitive abilities. The major shortcoming of strategic safety planning and risk assessment is that the available models are based on the presumptive assumptions of having infinite time to perform the analyses and a boundless amount of readily accessible data - both are unrealistic assumptions when dealing with dynamic large-scale systems. For example, several attempts to assign designated routes to hazardous material shipments have resulted in opposition by the transportation industry and in problems with reaching a regional consensus among governmental organizations representing the varying interests of affected communities.

Emergency response, as a post-event measure, is probably the most well organized and effective action in most of the LP-HC event domains. Even so, a report by the Office of Technology Assessment (OTA) concluded ten years ago that nearly 75% of the nation's police and firefighters are inadequately trained to respond to accidents involving hazardous materials [Haddow, 1987].

Concentrating efforts on improving only these traditional pre-event and post-event approaches will not result in satisfactory safety condition for operational activities. The most crucial shortcoming of these approaches is that they focus almost solely on the planning phase before a potentially hazardous event (pre-event measures) and on the organizational and technical tasks to be completed after an incident (post-event measures). For the duration of the operation itself (i.e., the time of a potentially catastrophic event), few effective safety approaches have been suggested, probably because the technology required for implementation has not been available.

It is our contention that the technology needed to provide safer operations is now available. Advanced computing and communications systems, coupled with appropriate reasoning models, will enable risk managers to close the gap between pre-event and post-event safety measures through real-time sensing and control of operational activities. This concept of operational risk management raises three basic questions.

- Why operational risk management?
- Is it technically feasible?
- What are the expected benefits?

The motivation for operational risk management will be discussed in the following sections, including a discussion of the generic tasks and the variables of operational risk management. The technological feasibility is discussed in terms of the availability of advanced communications technology, mainly mobile communications. The chapter ends with early reports on the expected benefits of employing operational management in general, with a special interpretation for operational risk management.

1.2 The Motivation for Operational Risk Management

The motivation for operational risk management emerged in the late 1980s in the field of hazardous material transport. Statistics reported by the U.S. Department of Transportation (DOT) showed then that there are more than 180 million shipments transporting 1.5 billion tons of hazardous materials each year in the United States, as illustrated in Figure I.1.

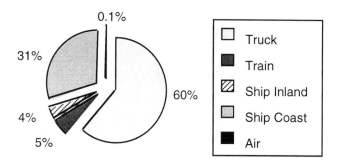

Figure I.1: Hazardous material transports of 1.5 tons per year [OTA, 1986].

The estimated amount of hazardous material shipments in the United States by DOT and the Corps of Engineers is 1.5 tons per year. What is remarkable is that 10% of all truck shipments involved hazardous materials and 5% of the air cargo at the 39 major airports contain (at least small amounts) of hazardous materials [OTA, 1986].

Shipments of hazardous materials are made by land, sea, and air. More than half of all hazardous material shipments (1 billion tons) are done by truck.

The vehicles used range from tank trucks, bulk cargo, carriers, and other specially designed mobile containers to conventional tractor trailers and flat beds that carry packages, cylinders, drums, and other small containers. 80 million tons a year (5% of all hazardous material shipments) are transported by rail, commonly in tank cars.

Most of the hazardous materials transported by barge on inland waterways is bulk cargo. The Corps of Engineers has estimated that the total inland waterborne volume is approximately 60 million tons per year. Coastal and inland waterborne volumes, combined, reach 550 million tons annually [OTA, 1986]. Transportation of hazardous materials by barge on inland waterways is a particular problem for regions with an extended water network connecting large urban centers.

The transport of hazardous materials by air is performed either in all-cargo aircraft or in the belly compartments of passenger aircraft. It is insignificant in total tonnage but constitutes a high number of shipments. A Federal Aviation Administration Study (1980, [OTA, 1986]) found that roughly 5% of the air cargo at 39 major airports (overall 300,000 packages) contained hazardous materials in rather small parcels of high-value or time-critical material [OTA, 1986].

Figure I.2 gives an overview of the regularly used modes for the annual 180 million shipments of hazardous materials in the United States (based on estimates made in 1982 [OTA, 1986].

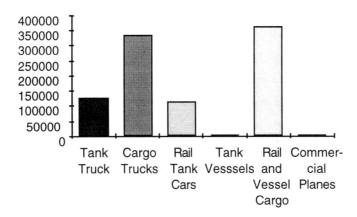

Figure I.2: Modes used for hazardous material shipments [OTA, 1986].

Data reported to the Office of Hazardous Material Transports (OHMT) by hazardous materials carriers showed an annual average of 1.25 incidents per 10,000 shipments for the period of 1973-83. Moreover, some experts estimated that there may be as many as three to four times as many incidents that go unreported. Statistics by DOT also showed that in the early 80's there was an annual average of 24 deaths and 663 injuries in hazardous material accidents which correspond to approximately one death or injury per 880,000 shipments [U.S. GAO, 1980].

Figure I.3 gives an overview of the incidents involving hazardous materials between 1973 and 1983 (as reported to DOT). The source is the U.S. Department of Transportation, Materials Transportation Bureau, Annual Report on Hazardous Material Transportation, Calendar Year 1983. The $13 million worth of damages is considered to be a conservative estimate. The actual figure might be as much as 10 times higher [OTA, 1986].

The Congressional Research Service's (CRS) study found that human error was judged to be the probable cause of 64% of incidents involving hazardous materials, followed by package failure (29%), vehicle accident/derailment (5%), and other (2%).

These statistics reflect annual averages and do not show the potential for catastrophic loss. A single accident causing catastrophic loss could change the statistics dramatically. For example, worst case estimates for LP-HC events showed that a major radioactive release in New York City could result in 3,000 deaths and decontamination costs of more than $2 billion, or that 18,000 city residents could be killed by an accident involving just one tanker of chlorine [U.S. GAO, 1980]. Several incidents in recent years have proved that the potential for catastrophic events cannot be ignored. The impact would have been of much greater consequence if some of the accidents happened at different times, in different places, or under different conditions.

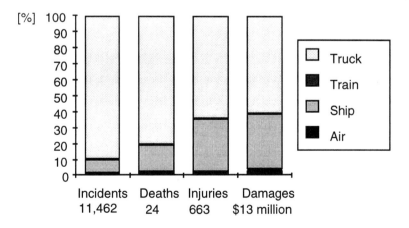

Figure I.3: Incidents involving hazardous materials [OTA, 1986].

Considering these statistics, the need for operational risk management can best be illustrated by looking at some spectacular accidents:

- In 1978, a train derailed in Waverly, Tennessee, causing a propane car to explode. 16 persons were killed and 50 were injured [Glickman, 1988].

  Train derailments are usually the result of tracks that have been badly designed or poorly maintained. The condition of the tracks and the environment could have been continuously monitored by a control center. Information on abnormal track conditions could have been forwarded to the engineers.

- In Kenner, Louisiana (November 1980), six persons were killed by a gasoline fireball created when a train collided with a gasoline tank truck at a grade crossing [Glickman, 1988].

  Operational risk management could have warned the truck driver of the approaching train.

- On December 3, 1985, a tank truck carrying 11,000 gallons of gasoline burst into flames as a result of a blown tire. Interstate 95 remained closed for several days, since one section of the highway had melted and needed to be replaced [Boghani and Mudan, 1987].

  Critical points on a vehicle, such as brakes and tires, could have been monitored as part of operational risk management. The driver could have been warned about the tire condition.

- In Littletown, New Hampshire, a tank trailer loaded with 9,200 gallons of propane jack-knifed on a icy hill and tipped on its side about 75 yards from a large storage tank of liquid propane and less than 100 yards from several large fuel oil storage tanks. Fortunately, no propane leaked from the truck, but a diesel fuel tank was ruptured. Until the propane was safely transferred, 1500 people within half a mile were evacuated (February 11, 1982) [Boghani and Mudan, 1987].

  A control unit could have continuously screened the road conditions and the environment in general. Here again, the driver could have been warned by the control unit.

- On July 11, 1978, a tank truck carrying liquid propylene ignited, causing a large fireball in the vicinity of a crowded campsite in Spain.

The catastrophic incident caused approximately 170 deaths and numerous injuries [Boghani and Mudan, 1987].

An operation center could have seen the truck approaching the campsite. An alternate route would have been computed and forwarded to the driver in order to avoid the densely populated area.

- At 11 PM on March 23, 1989, the captain of the EXXON Valdez radioed to the Coast Guard that he was turning left from the outbound to the inbound lane to avoid ice. Soon, the Coast Guard lost the ship on the radar. The captain turned over control to the unqualified third mate. The ship ran aground after failing to turn on time. Companies that boasted they had the equipment and manpower in place for a quick cleanup turned out to have hardly anything available and lost irreplaceable days getting ready to take action. The result was a 10.5 million oil spill, killing at least 34,000 birds and 984 sea otters - the largest spill in US history [Behar, 1989], [Witteman, 1989].

A satellite based operational risk management system would have been able to track the ship and the unqualified mate would have been identified by the control unit [Harrald et al., 1990].

These are some of the more spectacular accidents involving hazardous materials. Less dramatic disruptions occur almost daily. For example in 1980, the Hazardous Materials Response Team of Houston's Fire Department responded to 297 calls involving hazardous materials, amounting to an average of nearly one call per day. The Coast Guard reported 6,700 oil spills during 1988, ten of which involved at least 100,000 gallons.

It is evident that control of the movement of trucks, trains, and ships carrying hazardous cargo has indisputable benefits. Such control should not be considered an alternative to traditional strategic safety planning efforts but rather an integral component of a novel approach to risk management.

The concept of operational risk management as proposed for transportation of hazardous materials can be applied to any other operational activity with the potential for LP-HC events. One particularly relevant area is the operational phase of emergency response, when unforeseen occurrences can disturb the operation, leading to reduced effectiveness and increased consequences.

The strategic process of emergency response typically involves the activation of one or more contingency plans. These plans are implemented by the establishment of a command and control center. The on-scene commander coordinates the activities while his or her staff gather data, do analysis, make decisions, and monitor their implementation. The response activities are dangerous and must be performed under time pressure.

Traditional risk management approaches for operations with the potential for LP-HC events, such as hazardous material transportation and emergency response, focus on either the planning prior to the beginning of the operation (pre-planning), or on being able to respond quickly if an accident or unexpected event occurs during the operation (post-event activities). The period of time when the operations are underway has not till now been considered "controllable," due primarily to insufficient communications technologies (see Figure I.4); these technologies are now available. Before discussing these technologies, however, operational control needs to be defined and discussed in the context of risk management.

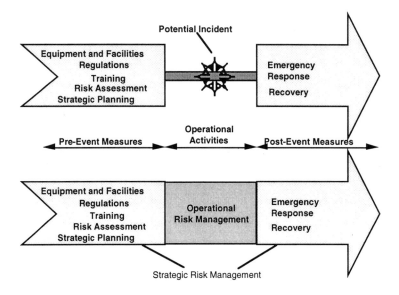

Figure I.4: Risk management approaches.

1.3 Operational Control: Definition, Tasks, and State of the Art

The process of **operational control** can be generalized as one of (1) sensing the present environment and perceiving an anticipated environment, (2) comparing both to a desired state, and (3) if there is a gap, taking action to bring the present or future state in concordance with the desired state. In order to effect this process, one must gather and store data. These data must then be processed in a form that is meaningful and of value to the decision making process. The activities of the control center, which will be referred to as **generic tasks**, can be subdivided into sensing and reasoning. **Sensing** refers to periodic checks, measurements, and environment surveillance. These tasks will be supported by weather channel, voice generators, satellite imagery, temperature and pressure recorders, etc. **Reasoning**, on the other hand, refers to analysis, risk assessment, simulation, routing and scheduling, and assistance to operating personnel. These tasks are supported by models, computation, heuristics, analogy, etc.

The **variables of control** can be divided into classes of flexibility. General parameters (usually constant for the operations phase) are the available technology to support the controllers and operators, regulations, and so on. More specific parameters include the material itself, origin and destination, and the transportation modes. More detailed parameters of control, which are highly sensitive to time, include the environment, the state of the material, and the condition of the operators (e.g., drivers and chemical response technicians). To illustrate the concept of operational control in the context of risk management, let's consider **transit control**, control of the movement of vehicles while in en route (Figure I.5).

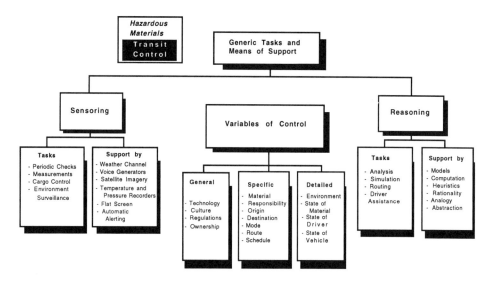

Figure I.5: Generic tasks and variables of transit control.

For the case of hazardous materials transport, there are three modes of operations: (1) land (truck, cars, train, etc.), (2) water (inland, offshore, open water), and (3) air. The control centers performing these tasks in a dynamic setting will be supported in sensing and reasoning by appropriate technologies.

These technologies include satellite imagery, decision support systems with advanced graphical interfaces, voice generators, and sensors, as addressed in the next section. The decision making tasks will be addressed through a human-machine system, where the technology will support the human making the decisions – the operational risk manager.

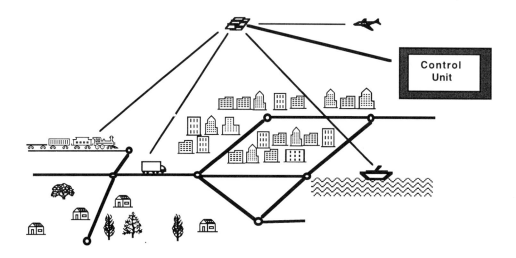

Figure I.6: Transit control of hazardous material flow on a network.

Safety in transit control is primarily concerned with two aspects. The first is to protect passengers using various modes of transportation; this is called internal safety. The second is to protect the environment surrounding the mode; this is called external safety. The term "safety" as used in regard to the shipment of hazardous materials usually refers to external safety. The safety of individuals who may be unaware of or uninvolved in the risky action of transporting hazardous materials is of major concern. Air traffic of hazardous materials would be the exception – one would be concerned with both. External safety seems to be the primary consideration for operational risk management in general and the transportation of hazardous materials in particular.

The most advanced organization in transit control is air traffic control. Safety aspects are directed primarily towards internal safety (passengers and cargo) rather than towards the external environment, largely because the confines of the external environment are difficult to define. Scheduling and routing are the most important control measures for public air transport. A similar situation exists in rail transport, where control has a significant historical basis. Early efforts to improve safety and productivity of railroad operations indicated the use of satellite imagery for control systems located in Rail Operations Control

Centers (ROCS) [Burns, 1989]. Safety was to be improved by (1) movement authority including monitoring, direction, and automatic enforcement, (2) monitoring and alerting of specified hazards, (3) continuous train position information, and (4) dispatcher-initiated emergency stops. Routing and scheduling are also major control measures for rail transport.

Latest developments in the maritime community include technological innovations from collision avoidance systems to satellite navigation systems. Despite the adequate technological advances, and the potential for effective regulation, the maritime culture has precluded any transit control. Therefore, there has not been any decline in the accident rate of about one ship loss per day; this rate carries the potential for an ecological disaster [Perrow, 1989]. Routing and scheduling are completely oriented towards productivity and all too often neglect aspects of safety.

The trucking industry in the late 1980s was the least developed transportation mode with respect to transit control. Some private companies had started to implement satellite based tracking systems, which permitted a dispatcher to control and route vehicles on-line [Morentz, 1989]. This technology gives the dispatcher real-time information about vehicle location, cargo condition (temperature), and gas tank level. Except for a few special cases in the nuclear industry, such transit control has not been used in the tracking of hazardous materials; however, results of a study concerned with the implementation of this technology for emergency response showed promising results, both from technical and economic viewpoints [Morentz, 1989]. In the trucking industry, schedules are typically set by demand and routing decisions made by the driver without external assistance.

Table I.1 shows an overview of the situation in transit control in the late 1980s for different modes, where "X" denotes where it is today used, and "(X)" denotes partial use.

Tasks / Modes	Safety internal	Safety external	Scheduling demand	Scheduling time	CA planning operator	CA planning manager
public air transport	X			X		X
private air transport	X		X		X (pilot)	
inland water		(X)			X (captain)	
off-shore water		(X)			X (captain)	
open water					X (captain)	
truck land transport	(X)	(X)			X (driver)	
train land transport	(X)	(X)				X

Table I.1: Employment of transit control in the late 1980s.

The objective of transit control is to improve external safety for the transport of hazardous materials without compromising the safety of the driver, crew, and cargo (internal safety). In order to achieve this objective, time driven scheduling and dispatcher controlled routing must be emphasized. The air and rail modes came closest to what was seen in the late 1980s as effective operational control.

Advances in communications and sensing technologies have made operational control a reality. Of particular importance is the ability to control and communicate with mobile, dispersed operations, such as trucks carrying hazardous materials and emergency response workers performing hazardous operations. Wireless communications technologies especially satellite systems, are the key advance.

2. Advanced Communications Technologies

2.1 Introduction

Satellite technology dates back to 1985 with the launch of Score, the first artificial satellite for voice communications. The U.S. Communications Satellite Act of 1962 which lead to the formation of the Comsat (Communications Satellite Corporation) was a milestone in satellite communications. It was the forerunner of the multinational organization Intelsat to which over 100 nations belong [Pritchard et al., 1993].

Six Transit satellites emitting two frequencies were launched into orbit between 1967 and 1973. With these two frequencies, distances to the receivers could be determined. GEOS-C satellite was another satellite-based surveying approach which became operational in 1975. Its major purpose was to perform altitude measures of the geoid. The U.S. Department of Defense (U.S. DoD) Global Positioning System (GPS) was the latest breakthrough in satellite-based positioning, surveying, and geodesy.

The use of commercial satellite communications systems started in the early eighties when the costs of satellite transmissions dropped drastically. Examples of major promoters of satellite communications and positioning systems are Intelsat (International Telecommunication Satellite Consortium), Inmarsat (International Maritime Satellite Consortium), the U.S. DoD with GPS, and Eutelsat (European Organization for Satellite Telecommunications).

Commercial satellite-based positioning and communications systems started to take off in the late 1980s and early 1990s. Governments were promoting satellite systems, manufacturers were developing special interfaces, service-providers were integrating systems and software, and prospective users were planning future applications of the new technology. However, many corporations, manufacturers, and service providers disappeared as fast as they appeared on the scene. While the technology had matured to a high level, the application field remained in its infancy for a long time.

Satellite technology plays an important role in navigation, both for data transmission (messaging) and for location of mobile units (positioning). Satellite communications can be employed for both tasks, although some of them could be accomplished with technology other than satellites. For example, positioning could be done with the long range area navigation system (Loran), and messaging could be done with a radio system.

All sorts of data, including binary signals, voice, text, graphics, and video, can be transmitted by satellite systems. A distinction is made between voice, video, and other kinds of data. The difference between voice, video and (other) data can be explained functionally rather than technologically. Voice and video communications can be received and processed by humans directly. Binary data must first be processed in order to be meaningful to humans.

Voice, video and binary signals can be received in either raw or processed. As mentioned above, most data must be processed in order to be meaningful to a decision maker; however, some data are not meant to be meaningful to a decision maker and thus do not need additional processing. For example, data can be used to trigger a machine or to indicate that an event has occurred (e.g., by turning on a warning light). Data that gets processed before it reaches the decision maker may be considered intelligent communications, as opposed to what might be called data unprocessed or "naive" communications. Intelligent systems using a variety of data can play an important role in intelligent communications. For example, voice recognition technology can use voice (or text) as input to a knowledge-based system, or data can be processed using knowledge-based technology.

Satellite and mobile communications share common characteristics, but they are not necessarily identical; that is, mobile communications systems need not be based on satellite technology. For example, mobile communications could be accomplished by digital cellular systems. Examples of satellite based mobile communications approaches are Motorola's Iridium system and Globalstar.

More important than the distinction between mobile and satellite communication is the difference between mobile and stationary communications. Electronic stationary communications became operational more than a century ago with the telegraph system. Mobile communications became necessary with the advent of aviation and later space exploration. The first technical communications media that were used for mobile communications were radio systems. Today, phones and satellite systems have given way to other viable alternatives. Mobile communications refers to a situation where at least one communications partner is moving during the communications process; for example, the communicant may be traveling on the road, on water, or in the air.

Stationary point-to-point communications can be delayed or performed in real-time. Telephone communication is real-time, while electronic mail communications is usually delayed. Satellite systems can support both real-time and delayed communications. However, the real benefit of satellite communications lies in its capability of providing real-time and mobile communications, independent of location and time of day, in order to provide the basis for monitoring, tracking, and navigation systems.

Communications systems may utilize either direct or processed communications. Direct communications between two units is effected directly via a data transmitter, for example, a satellite system. Satellite communications providers can also process data at a central data processing station, which determines which data goes to which client, usually on a store-and-forward basis. Examples of satellite communications systems working with this principle are the EutelTracs and the Inmarsat C systems.

Satellite communications systems may be either open or closed systems. Open communications systems separate transmission from positioning communications, while closed systems do not. The EutelTracs system used in Europe for navigation is a closed system that provides both data transmission and positioning. But many satellite-based monitoring systems use GPS for positioning of mobile units and Inmarsat C for transferring position data to headquarters.

Although satellite communications can support both local and global areas, satellite communications systems are seldom used in local communications. Technological alternatives also exist for wide area communications, particularly with the networking infrastructure provided by the Internet. Satellite communications is especially valuable for wide area, long distance, and mobile communications purposes.

Closed satellite communications systems transfer data to earth stations and from there via terrestrial networks to processing and distribution centers. Links between communications and information systems must be addressed at both the functional and technological levels. The functional level consists of the tasks that can be performed with satellite communications systems, including monitoring and decision support. The technological level refers to the control of the satellite system and to network management, including performance assessment, fault diagnosis, and network design. These aspects of telecommunications will be addressed in this chapter.

Satellite communications technology is a fast growing technology, providing the basis for many new management and business approaches. The vast amount of data that can be transferred must be processed by intelligent systems. Knowledge-based technology has several direct and indirect links to information and satellite communications technology, including geographic information systems and multimedia systems.

2.2 Overview of Satellite Communications Technology

The **Inmarsat** partnership is a cooperation between commercial companies, government authorities, and telecommunications organizations around the world. They include Inmarsat, Inmarsat signatories, land and earth stations (LES) operators, national communications authorities, equipment manufacturers, and value-added service providers. Inmarsat is a cooperative, commercially-oriented enterprise, with its headquarters in London (UK) and more than 60 member countries. Its prime task is to establish, maintain, and operate the satellite system required to provide global mobile communications for maritime, aeronautical, and mobile users.

The Inmarsat C system offers two-way message communications between fixed telecommunications installations and mobile units around the world and around the clock. The length of the messages can be several pages. Communications in both directions are handled on a store-and-forward basis which means that messages are sent as a number of packets of digital data. After the last packet has been received via the satellite system, it takes a few minutes to reassemble the complete message. However, the store-and-forward system has the advantages of reducing communications costs and providing high flexibility in communications; for example, using telex, fax, e-mail, and switched data.

Inmarsat C has a number of service capabilities which include polling and data reporting, position reporting, safety/emergency alerting, anti-theft/hijacking, and remote sensing and control. In addition, fleet management and dispatching of safety warnings can also be performed with Inmarsat C.

Inmarsat M is a low-cost voice satellite communications system. It allows two-way voice, optical fax, and data communications, and its interface can be integrated into a telephone key-pad. The mobile unit fits in a regular suitcase and weighs less than 30 pounds, so that it is readily portable. Although satellite phone communications provides worldwide service, the speech quality can be poorer than that of terrestrial mobile system. Special features have been built into

Inmarsat M systems. These include remote mounted distress buttons that connect automatically to rescue authorities; alternative language voice prompts for operator convenience; and special function keys.

The **Global Positioning System** (GPS) is a satellite system developed by the U.S. Department of Defense which is based on 24 Navstar satellites orbiting the earth at an altitude of 10,900 nautical miles. Because of their high altitude, their orbits are very predictable. Minor variations in orbits are measured constantly by the DoD and data is transmitted from the satellites themselves. The expected life span of a satellite is 7.5 years.

The accuracy of GPS locations is determined by several sources of error, where the contribution of each source may vary depending on atmospheric and equipment conditions. In addition, GPS accuracy can purposefully be degraded by the DoD using an operational mode called selective availability (S/A). The S/A mode is designed to deny hostile units the advantage of GPS positioning. When it is implemented, it is the largest source of error. A typical good receiver has an accuracy of about 20-30 meters, in the worst-case about 60 meters. When S/A is implemented, the accuracy reduces to about 120 meters.

This accuracy might be sufficient for many purposes. However, surveying and positioning tasks require higher accuracy, which can be achieved with differential GPS (dGPS). With dGPS, a receiver which is placed at a known location calculates the combined error in the satellite range data. This correction can be applied in the same locale, eliminating basically all errors in their measurements, to obtain a precision better than a meter.

GPS provides accurate positioning and navigation around the clock and at any place on earth. Therefore, GPS is used for a multitude of tasks, including surveying, vehicle tracking, air collision avoidance, and zero-visibility landing assistance. However, GPS position data are received at the mobile units and might have to be transferred to the headquarters for further processing. In order to transmit these data to the company's headquarters, a communications system must be connected to the GPS system, such as the Inmarsat M system.

The European organization for satellite telecommunications, **Eutelsat**, was founded in 1977 by the national post-telephone-telegraph (PTT) consortiums of 17 European countries. It is being managed by an intergovernmental convention established in 1985 and is open to all European countries. The organization, with its headquarters in Paris, counts 48 member countries [Grenier, 1996].

The purpose of Eutelsat is to provide satellites to satisfy the demand for public services in telecommunications in Europe. These include telephone, telegraph, telex, fax, any sort of data communications, radio and television broadcasts, and mobile terrestrial services. In addition, all services necessary to maintain the satellites are also provided.

Since 1990, Eutelsat provides EutelTracs, a mobile terrestrial satellite positioning and communications system. From company headquarters, transportation managers can monitor on a computer screen the location and movement of their vehicle fleet. The conceptual layout of the EutelTracs system is shown in Figure I.7. It consists of two Eutelsat satellites: one for messaging and the other for positioning. Messaging includes data communications, and positioning consists of measuring the location of the mobile and/or remote units via triangulation.

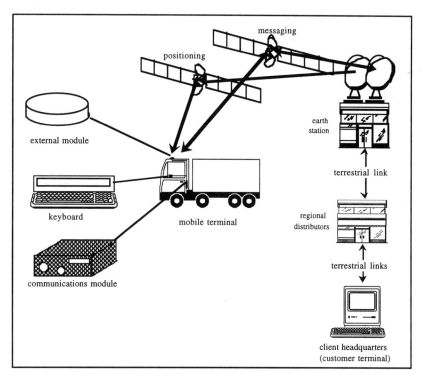

Figure I.7: Overview of the EutelTracs system.

The analog system to EutelTracs in the U.S. is OmniTracs; the U.S. also uses two satellites that cover different areas on the earth's surface (North America and parts of Central America). The two satellites are in geostationary orbits at a height of 35,786 km, and are controlled from a control hub station which has two antennas. From there, data are transferred to the service providers, which forward and receive data to and from the company's headquarters.

The equipment needed to work with the EutelTracs system consists of a micro-computer with appropriate software at headquarters, and, for each mobile unit, an external module (attached at the roof of the mobile unit that contains an antenna, an amplifier, and converters), a keyboard (with a display of four lines with forty characters each), and a communications module (32x23x11 cm).

The service area of EutelTracs includes Europe, the northern part of Africa, and parts of the Middle East. The services provided by the EutelTracs system include two-way communications between headquarters and mobile units; transfer of up to 2,000 characters per message; predefined macro messages to be sent to mobile units; positioning of mobile units with 500 meters precision; and tracking of mobile units on geographic background.

2.3 Applications of Communications Technologies

Potential applications for satellite communications are:

- those for which <u>distance</u> is an issue (for use in voice/data/video communications),
- those involving <u>real-time</u> communications (for use in on-board positioning), and
- those for which <u>mobility</u> is a requirement (for use in transportation).

Many tasks supported by satellite communications combine at least two of these three characteristics. Figure I.8 shows examples of tasks where satellite communications can play an important role.

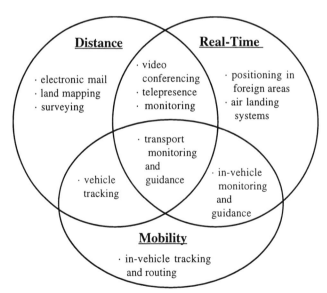

Figure I.8: Characteristics of potential applications of satellite communications.

Some tasks refer solely to real-time aspects; that is, tasks which do not need mobile communications over long distances. This kind of satellite communications includes all on-site positioning tasks, such as a person determining his/her position with a satellite receiver in real-time or an airplane landing with GPS-based zero visibility landing systems.

Satellite communications also plays a major role when long distances are the primary concern; that is, stationary tasks which are not performed in real-time. Examples include electronic mail systems and all surveying tasks where the data are transferred directly to the information systems at headquarters. Local and asynchronous mobile tasks (i.e., short distance and not in real-time) can rely on technologies other than satellite communications. Examples are in-vehicle tracking and routing. The system checks where the vehicle has passed and determines changes in the route.

Tasks characterized by combinations of distance, real-time, and mobility are most appropriate for satellite technology. Real-time distance tasks can be performed by television and telepresence using virtual reality technology, where a person operates a robot from a distant and safe place (see Chapter III). Monitoring of stationary systems or of environmental regions and weather conditions are additional examples of distant real-time tasks. Examples of distant and mobile satellite communications include vehicle tracking and navigation tasks, where managers at a central headquarters can check (not in real-time) the routes of the vehicles and use this information to see which customers have already been served by the delivery trucks. Real-time and mobile satellite communications can be used for in-vehicle monitoring and guidance. The driver, pilot, or captain monitors the vehicles' current position and determines changes in the current position based on this real-time information. Satellite communications technology is also appropriate for those tasks where all three attributes hold. Examples of mobile, distant, and real-time satellite communications tasks include vehicle monitoring and guidance from a centralized operations center, for road, water, and air transportation.

The growth and impact of the Internet is expected to be greater than the growth and impact of the personal computer, creating new opportunities for organizations and companies (e.g., advertising, virtual malls, information services) accompanied by large savings in information systems, systems development, and operation costs [Roche, 1995]. The number of hosts has grown from two millions in 1993 to 20 millions in 1997 [Lange, 1998]. Communications links allow companies to become virtual organizations by linking to other companies. Virtual organizations have the advantages of sharing costs and skills and of enhancing access to global markets; the disadvantages are a potential loss of operational and strategic control, coupled with the need for new management approaches [Dess et al., 1996]. Despite these promising perspectives, two issues need special considerations in world-wide communications networks: interoperability and security.

Interoperability refers to the ability of two or more systems in order to interact to achieve a predictable result. The liberalization of the communications market leads naturally to competing service providers with interoperability problems. An example of missing interoperability is the inability to exchange data or to communicate between different software applications, such as geographic information systems and hypermedia systems. This kind of software interoperability is essential because managers tend more and more to adapt applications to their specific needs.

Security has become a major concern in world-wide networks. It is estimated that 80% of unauthorized computer accesses occur via the Internet [Roche, 1995]. Encryption, passwords, and "fire-walls" are commonly used prevention and mitigation measures. However, non-malevolent disruptions of the communications and information processing system can cause harm to an organization. For example, management must have contingency plans for when the electronic vehicle tracking and communications system goes down.

Two major trends are developing in business communications: multimedia personal computers and wireless communications. The first has the potential for integrating intelligent systems, while the second provides the basis for satellite-based video, text, audio, and any other data communications. New ways to communicate, in addition to audio and data conferencing, include group video conferencing, desktop video conferencing and interactive graphics teleconferencing. In 1995, 60,000 to 100,000 people were using personal desktop conferencing systems; this number will increase exponentially until the end of the decade [Grierson, 1996]. Desktop video conferencing allows managers to work closely together on time critical projects and share and exchange data and information. Desktop and general video conferencing will also affect the interpersonal communications structure of organizations. Managers do not have to gather in one place to meet for internal business meetings and employees can perform their jobs at home or from any other work place. This latter development, referred to as telecommuting, can cut down the administrative and logistic burden of an organization and at the same time increase flexibility.

Another significant development concerning intelligent satellite communications is intelligent agent technology (IAT). An intelligent agent is an autonomous piece of software that can traverse a network to accomplish tasks remotely and act independently when it encounters problems [Keyes, 1996]. Keyes [1996] identifies five characteristics that an intelligent agent must be capable of: running on any type of computer or operating system; gathering information; making a decision based on the answers to a series of questions; acting on that decision or aborting the task; and reporting the results. Tasks that intelligent agents can address include software upgrading in a network (the intelligent agent identifies which machines around the world need to be upgraded and how); and data collection across the world (the manager can specify which type of data is wanted and how to prevent overload) [Keyes, 1996].

2.4 Outlook

Satellite communications technology has grown immensely over the last two decades. While the pioneer work was conducted by a few national governments, multinational consortiums have evolved, like Eutelsat, Inmarsat and Intelsat. Satellite density is growing so fast that it has become a major concern in the satellite communications field. Small home terminals for direct broadcast and mobile satellite applications are expected to number in the hundreds of millions and the growth rates for the satellite communications industry are estimated to be over 100% per year [Pritchard et al., 1993].

The quality of voice and video communications is improving steadily. This, in combination with advanced information technology, including knowledge-based systems, multimedia, and virtual reality technologies, provides the basis for further exploitation of the role of decision support systems in real-time satellite communications.

The number of wireless subscribers to cell telephones in the U.S. has grown from 5 millions in 1991 to almost 50 millions in 1997; the growth is even more impressive for remote regions, as in Scandinavia where almost every other person has a mobile phone [Riezenman, 1998]. Technological innovations have no borders, as Motorola's Iridium and the Globalstar systems clearly show. Iridium consists of 66 LEO satellites, providing mobile satellite service for handheld personal telephones – world-wide [Schneiderman, 1994]. The Globalstar system will offer wireless telephone and other digital telecommunications services, including data transmission, paging, facsimile and position location. Full operation of the 48 satellite system is expected for 1999.

There is no doubt that technology can support these advances in satellite communications. Organizations are globalizing, promoting distributed services and work places. Remote and mobile activities can be monitored, controlled, and supported in real-time. As a result, concepts like the intelligent transport system, tele-conferencing, long-distance virtual reality applications, virtual organizations, intelligent agent technology, and telecommuting are being developed and implemented.

To complement the technological development, supporting policies must be devised and enacted. A major prerequisite for world-wide mobile satellite communications is the use of equipment in different countries. However, there might be national requirements for using satellite equipment from other countries. This problem is referred to as transborder operation. The European Community recognized this problem in the early stages of satellite communications and introduced a circulation card which allows the use of satellite equipment within Europe.

Deregulation of telecommunications services is another major issue that affects the commercial use of real-time satellite communications technology. The abandonment of the 50-50 rule for sharing traffic between satellites and cables over the Atlantic by the U.S. federal communications commission (FCC) promotes the implementation of privately owned satellite systems, in competition with Intelsat [Tirró, 1993, p. 6].

In 1990, the European Space Agency (ESA) established a working group on satellite communications policy for the purpose of drafting ESA's response to the proposal contained in the Green Paper of the Commission of the European Communities (CEC) on satellite communications [COM(90)490, 1990]. The results of this deliberation are reported in [ESA, 1993]. Among the recommendations were (1) that the EC press on with its measures to liberalize the international market; (2) that industry focus its restructuring efforts on the capability to offer and to operate complete customized turn-key systems; and (3) that national telecommunications operators either create subsidiaries fully dedicated to satellite services or leave the market entirely to private enterprises.

The managerial issues that arise from these developments in satellite communications and information technologies evolve along with the continuous growth of the technologies. In general, the managerial issues can be divided into four classes: technology, operation, organization, and policy making.

Technological issues refer to the development of satellite communications and information technologies, including knowledge-based systems, multimedia, and virtual reality concepts. The steady growth of these systems will lead to advanced intelligent networks. As a result, managers will embrace a large set of new services that can be integrated into their operations. Examples are world-wide mobile communications, EDI, multimedia electronic newspapers, PC-based video conferencing, fraud prevention and information systems, and so on.

Operational issues refer to the use of satellite technology and the integration of knowledge-based systems technology in support of human decision making. Systems will be developed with the potential for interoperability and with open architecture. Interfaces between the technology and the user, including the definition of novel tasks that can be addressed with the new technology, will become easier to build. The increase in interoperability will help managers to "assemble" their specific applications within and across commercial software and communications systems.

Organizational issues refer to reengineering of processes due to the advancement of real-time distributed management, global logistics for inventory control of remote distribution centers, and real-time management of communications and processes on a world-wide basis. Special attention must be given to the transition from traditional tasks to new tasks that are being addressed with the new technology. It is important for management to realize that any new technology must be carefully integrated in new work processes, and that operators need time to adapt to these new environments.

Finally, organizational, national, and global policy issues must be considered in light of the desire to promote further growth of the technology and its appropriate use. The rapid growth of technological developments and the commercialization of satellite communications and knowledge-based systems technologies overwhelm the market with revolutionary communications and information processing products. Service providers produce useful systems, developers tailor these systems to the users' needs, and scientists assess their potential benefits and shortcomings in order to improve the products. In order to successfully promote these developments, however, a receptive environment must be cultivated. This includes a technology friendly attitude from the managements of businesses of all size, national services supporting the use of these new technologies, and international regulations promoting the use of satellite and information processing technologies.

Advances in communications technology have made possible the control of operations with the potential for LP-HC events. Most of these operations, such as transportation of hazardous materials and emergency response, might be dispersed over great distances, requiring real-time control and operational risk management. Wireless, potentially satellite, communications provides the technology to support management of these operations. Recent changes in governmental regulations, coupled with global demand from both industry and government are creating markets for the technology needed for the application of satellite communications. There is still, however, the need to support the decision maker, in this case the operations manager, in making effective use of the technology. To do so, the data provided by the technology must be processed into information useful in decision making. The use of models with appropriate human-computer interfaces is one means of providing decision support. Operational risk management, a new paradigm for operational control of hazardous operations, provides a logic that can be used to construct these models – and advanced multimedia technology can provide the interface.

3. The Concept of Operational Risk Management

3.1 Basic Concept

Large-scale operations involving low probability, sudden onset events with the potential for catastrophic impact are typically managed by well formulated, predetermined procedures. These procedures (also called courses of action) are designed to ensure that the operations are as safe and as cost-effective as possible. Unfortunately, large-scale operations are exposed to episodic events that can threaten the successful completion of a planned course of action. Real-time decision making is needed to respond to these unexpected events in order to prevent or mitigate undesirable consequences - in terms of both safety and costs.

Operational decision making has been addressed for various large-scale operational systems, such as air traffic control [Schutte et al., 1987], vehicle navigation [Karimi and Krakiwsky, 1988], mission planning [Beaton et al., 1987], and process control [Naum et al., 1989], [Moore et al., 1988]. The methods proposed typically focus on the selection and implementation of planned activities. They do not address the need for revision of these procedures in the light of unanticipated, episodic events. As noted in a handbook for chemical process safety, "performing high quality hazard evaluations throughout the lifetime of a process cannot guarantee that accidents will not occur" [CCPS, 1992]. In addition, emergency procedures themselves are susceptible to unforeseen events; for example, a "planned for" emergency vehicle may not be operational when needed. Dealing with the unexpected is still the province of the human decision maker. Therefore, technological aids for supporting decision makers in these typically stressful environments must consider the cognitive limitations of humans in dealing with safety and cost issues in real-time decision making.

Sudden onset events do not necessarily have to affect present operations. A truck carrying hazardous materials could, for example, be scheduled to drive through a region where a snow storm is expected or already in progress. If the dispatcher could "see and think ahead," s/he could assess the impact of such an event on the planned route and evaluate alternate routes; that is, avoid the snow storm by changing the planned course of action. Another example is specialized equipment that is to be used in a final stage of an emergency response, but is found to be unavailable. If the on-scene manager could "see and think ahead," that person could change the planned response to a course of action that does not need that particular equipment.

However, "seeing and thinking ahead," which is the sensing and reasoning needed for operational risk management and decision making, has only just become possible, because the technology for monitoring the environment for critical events and communications technology to support real-time reasoning has just become available. As noted in the preceding section, the latest advances in satellite positioning technology and mobile communication make real-time support for hazardous operations a feasible approach. These advances permit operators to "see" inside a damaged pressure vessel with micro-video, "feel" the conditions inside a burning container with robot devices, and "track" the dispersion of hazardous materials.

Advanced communications and information technologies provide new capabilities for monitoring and controlling large-scale operations. Operational risk management is capitalizing on these advances so that pre-planned courses of action can be revised in real-time whenever a critical event occurs. Although advanced technology can automate many tasks, humans will always be an integral part of managing large-scale operations and will require decision support in responding to low probability episodic events with the possibility of catastrophic impact.

Operational risk management can be defined as the set of activities designed to control several operations posing a potential thread to the system and the environment. The control center accomplishing the tasks of operational risk management will be able to manage simultaneously several hundred discrete hazardous operations. It will be equipped with several computer terminals for human control, a flat display for projection of the transportation network and visual monitoring of the hazardous operations, and voice alerting. Incoming data from periodical sensing of the sources of risk must be filtered and processed. Reasoning and data management are supported by knowledge-based technology, requiring human intervention only in cases of increased alert. Any significant change in the risk sources will appear on the large display, for minor events in yellow, and for major events in red. Besides the control tasks, other secondary activities can be performed in the control center, such as simulation of unfavorable conditions, display of potential dispersion clouds, or even simulation and testing of emergency response plans. Such control centers will be staffed on a 24 hour basis with a small group of operational risk managers.

The real-time environment for managing large-scale systems involving hazardous operations consists of three major components (see Figure I.9): (1) The large-scale operational system where operations take place and where events can occur that may affect the planned course of action; (2) the real-time controller who monitors the operation for accurate and timely performance and the potentially critical events, and reasons about whether the course of action should be revised; and (3) the communication links for data transfer between the system and the real-time controller. Due to advanced communication technology, the real-time controller does not have to be located with the operational system. It is assumed, though, that the loop between sensing and reasoning will always include humans.

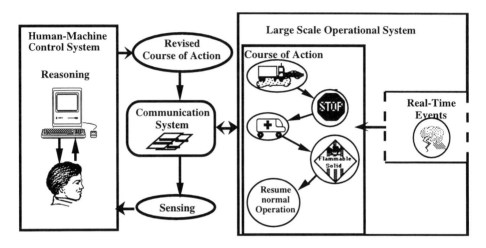

Figure I.9: Environment for operational risk management.

Real-time events (RTEs) are occurrences which may force a change in the present course of action. RTEs are unpredictable and occur without warning. Therefore, it is crucial that the human operator sense all RTEs. To do this s/he must continuously monitor the system and its environment. Both the need to reevaluate a course of action and the time available for sensing and reasoning are event driven. A reevaluation is necessary if at least one of the monitored operations is threatened by the RTE (e.g., the fire truck is useless for response if it is out of order). The time available to make decisions on a new course of action depends upon when the impact of the RTE affects the present course of action. Since the present course of action is known, this time interval can be estimated. To ensure safe and cost-effective operations of large-scale systems, not only must the appropriate course of action be selected but deadlines must be met.

RTEs seldom occur but when they do occur, the operational risk manager must have quick and efficient support in risk assessment and decision making. RTEs can last for a few hours up to a few days. Six major risk classes of RTEs can be delineated: (1) natural phenomena (weather, earthquake, etc.), (2) characteristics of operators; that is, those responsible for controlling the operation, such as vehicles, vessel or airplane in transit, (3) exposure, such as the population along a transport route or in the neighborhood of an industrial area, or ground water areas, (4) operation, such as the condition of the vehicle or ship (tires, brakes, steering, etc.), (5) cargo (temperature, pressure, etc.), and (6) infrastructure, such as the condition of the road surface and the traffic. Table I.2 shows the different classes for real-time events.

Classes for Real-Time Events		
Natural Phenomena – Weather – Earthquake	**Operator** – Condition of Driver	**Exposure** – Population – Environment
Operation – Tires – Brakes – Tank	**Cargo** – Temperature – Pressure	**Infrastructure** – Road Surface – Traffic

Table I.2: Classes for real-time events.

In Figure I.9, Table I.2, and the discussion to follow, the transportation of hazardous materials is used to describe the concept of operational risk management. Other situations, like emergency response, will be used to illustrate the generality of the concept and its applicability to management of any operations subject to low probability, sudden onset events with the potential for consequential impacts on life, property, or the environment.

Given the continuous update of real-time data and information, the tasks and the decision processes pertaining to monitoring, risk/cost assessment, and selection or courses of actions must be analyzed from a dynamic point of view. The operational risk manager changes from the monitoring task to the risk/cost assessment task only if an RTE seems to affect safety or operational costs of monitored operations. The occurrence of RTEs, such as bad weather or traffic accidents, is unpredictable. Data and information about the event can be very limited both in quantity and quality. The first task is therefore to filter and process any incoming information. Depending on the results, the operation manager might decide to alert all operations immediately or only those that may be affected by the RTE. An operation is affected by an RTE if one or more aspects of its planned actions are affected by the RTE. For example, if the planned route of a hazardous material shipment goes through a bad weather area, the shipment would be considered as affected in its operation.

One important parameter is the time it takes for operations to be affected by the RTE. For example, if the hazardous material shipment is far away from the region affected by a storm, the dispatcher can postpone his rerouting decisions, while he must act fast for those shipments which are close or already within the region.

After risks and costs of the affected activities have been assessed, a decision support system at the control center would compute new optimal courses of action for the affected operations, generating three types of results. First, it can suggest stopping certain operations (e.g., hazardous material shipments) and waiting until the RTE (e.g., snow storm) is over, second, it can suggest resequencing certain operations to avoid hazardous operational conditions (e.g., reroute the shipments to avoid the snow storm), and third, it can suggest continuing the operation as planned if the RTE is not too dangerous. The updated courses of actions are then transferred by the operational risk managers at the headquarters to the operators.

In general, the operational risk manager responsible for the operations must sense the environment and reason about how to respond to unexpected changes. The resequenced decisions are then implemented by the operating units (e.g., the drivers). In the case of transportation, the first aspect of sensing refers to obtaining requests from customers. This can be accomplished by questioning electronic cargo bourses or by monitoring the customers' needs. The second aspect of sensing is to monitor the locations of the operations and their status, preferably on a large monitor or screen. Communications between the operations center and the operations (e.g., vehicle) could then be done by clicking with the computer mouse on the appropriate operation on a large screen (e.g., a vehicle on a map background). The third aspect of sensing is to monitor the environment for any changes that could improve or worsen the on-going operations.

An operation (e.g., shipment) is considered to be successful if it is completed on time and safely. Timeliness refers to customer satisfaction and economic gains. Safety refers to internal safety of the drivers and the cargo as well as to external safety of the population and the environment. External safety is especially crucial for hazardous material shipments. The quality of any shipment deteriorates if its safety is reduced or if delays occur. The quality of a shipment improves if sudden changes enhance profitability. For example, if the forecast of bad weather turns out to be wrong, both safety and timeliness (and thus the overall quality) of the shipment improve.

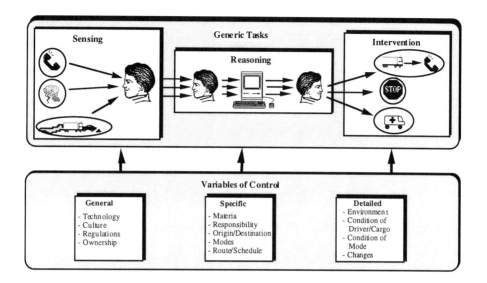

Figure I.10: Generic tasks and variables of control for operational risk management.

Reasoning amounts to analyzing data and making decisions in a time constrained environment under uncertainty. Incoming data are often unreliable and incomplete. Managers at the operations center must make decisions while the operations keep progressing. Reasoning for transportation addresses three major tasks. First, orders must be assigned to a vehicle in such a way as to minimize transportation time and reduce empty shipments. Second, changes of operations in response to sudden onset events, which either threaten the successful completion of the shipment or promise more gains, must be evaluated and forwarded to the drivers. And finally, actions to effectively and rapidly respond to vehicle failures, such as breakdowns, must be determined and forwarded to the appropriate response unit.

The decisions made by the operations center are implemented either on the vehicles by changing their routes and priorities or by external units that rush to help the vehicles and the cargo. If the proposed decisions cannot be implemented, the operations center must investigate alternative courses of action.

Sensing and reasoning are not necessarily done by only one decision maker or risk analyst. Figure I.11 shows different decision situations, including automatic control (no human involved); individual analyst, where decisions are made by one person without decision support; multi-participant situations with multiple-experts in a face-to-face setting; multiple-experts in a distributed setting; and multiple experts in a cooperative setting.

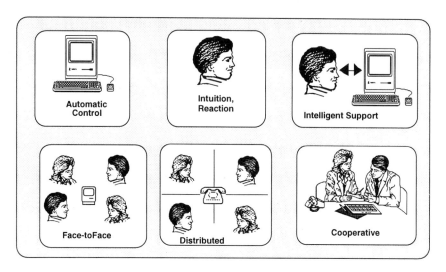

Figure I.11: Different approaches to sensing and reasoning.

In following chapters, a logic will be presented that provides decision support for the individual as well as for multiple expert decision makers. Operational risk management consists of a continuous monitoring and assessment of (1) the large scale system and its environment; (2) a predetermined set of activities and operations; and (3) the performance of these activities and operations. For example, consider an emergency manager who is monitoring a response operation where contaminant of a hazardous material spill is underway in a moderate wind. If the wind changes suddenly, the emergency manager must consider a new course of action. Such decisions to change a predetermined set of activities are typically based on the managers' experience and intuition. Providing decision support to operational risk managers used to be difficult because the technology for real-time monitoring and effective communications between operating centers and remote operations personnel was not available. In addition, the strategic or predetermined courses of action can only be of limited assistance, since the managers must deal with unexpected events that can occur suddenly and may have disastrous consequences - events that are difficult to plan for strategically.

As previously noted, traditional approaches to risk management focus on either prevent or mitigation activities that are considered before the sudden onset event occurs, or response and recovery activities that take place after the event has occurred. The period of time that the operation is underway typically is not addressed in terms of risk management. The operational risk manager, or dispatcher, plant operator, or emergency manager, has only the courses of action that are determined by a plan; for example, routes, schedules, or emergency response plans. S/he therefore must rely on experience and intuition to deal with events that have not been addressed by strategic planning.

Operational risk management is not meant to be an alternative to strategic planning. Rather, it should complement strategic risk management for issues that cannot be addressed from a planning perspective. Consequently, any analytic approach proposed for operational risk management must be in accordance with strategic considerations, and vice versa. However, there are some fundamental differences in the motivations for strategic and operational risk management, as summarized in Table I.3.

Strategic Risk Management		**Operational Risk Management**
planning	↔	acting
before and after the operation	↔	during the operation
thorough data collection	↔	rapid data transfer
sufficient time for decision making	↔	limited time for decision making
sufficient estimates	↔	high uncertainty
probability of events matters	↔	possibility of events matters
policy decision making	↔	designation of courses of action
outcome feedback	↔	cognitive feedback
technology poses constraints	↔	human poses constraints
rational decision making	↔	emotional decision making

Table I.3: Differences between strategic and operational risk management.

3.2 Practicality of Operational Risk Management for Transportation Systems

The most significant technological innovation for supporting operational risk management in transportation is satellite-based vehicle identification and location coupled with on-line communication between the operations and headquarters. For example, vehicle location by satellite can be performed with sufficient accuracy for many needs [Briskman, 1988]. This technology enables a dispatch unit to monitor and control the actual location of any vehicle in real-time – at any time over a large geographical area. The ability to locate vehicles by satellite has drawn considerable attention due to its favorable cost/performance rate. Transportation carriers use this technology because it can offer an economic advantage in timeliness and energy costs. However, little attention has been paid to public safety attributes of these tracking systems. Many different vehicle location navigation systems using satellites are in operation, being implemented, or being proposed. Their performance varies according to their characteristics and user requirements. The gamut of vehicle navigation systems ranges from radio based navigation systems to voice recognition and generation; the latter will be available in the near future. For radio navigation, private or public radio stations transmit updated traffic situations and provide the driver with useful information about actual traffic conditions.

Several European countries (i.e., West Germany, Austria, Switzerland, and Luxembourg) started during the early 1970's to implement a radio information system for motorists [Buchman, 1988]. Traffic police cars transmit messages to the police operation center which are then passed on to the traffic studios of the various regional FM-broadcast stations. From there, the traffic information is transmitted on a special frequency. Car radios are equipped with a decoder that causes the special radio frequency to automatically turn on the radio or switch to the traffic radio messages, regardless of the radio's actual operational mode (e.g., tape, zero speaker volume, radio off, etc.). These messages not only describe the traffic situation but also provide motorists with navigation recommendations such as detours [Buchman, 1988]. The usefulness of driver support, particularly for the shipment of hazardous materials, should not be underestimated, especially in Europe where a motorist has to drive only a few hundred miles before entering a new country, with a foreign language and different traffic regulations (e.g., no night driving).

PROMETHEUS was a major research program in vehicle tracking involving the European car industry, the electronics industry, and universities. It included research on custom hardware for intelligent processing in vehicles, methods and standards for communication, and traffic scenario for assessment and new systems. Industry research is oriented toward driver assistance by electronic systems, vehicle-to-vehicle communications, and vehicle to environment communications.

The most commonly used radio-based location system in the U.S. was based on the commercially available LORAN-C transmission system used by the U.S. Coast Guard. LORAN-C provides continuous positioning to an accuracy of several tenths of a mile on a global basis. However, the Global Position System (GPS) has now emerged to the most relied on system for navigation.

Despite these advancements, navigation, as part of transit control, requires that appropriate interfaces and decision models interact with these satellite tracking systems. This requirement has been recognized in the emergency management community where efforts are underway to couple knowledge-based technology with mathematical models and integrate the result with an advanced geographical information system (GIS). GISs are ideally suited for capturing, analyzing, and displaying large volumes of land-based data in changing demographic and environmental conditions.

Most powerful GISs feature map digitizing, data transfer, relational database management, map overlay, display, query, interactive graphics display, network analysis, and multimedia capabilities. The most important feature for hazardous material transportation is the capability of geographic network analysis. GISs have the ability to analyze and model networks such as city streets, waterways, or highways. In addition, they can serve as a tool for vehicle route selection, and time/distance flow analysis.

In the late 1980s, the U.S. transportation industry began to implement commercially available satellite tracking systems based on OmniTracs, Geostar, and GPS. The motivation was to improve economic benefits and competitiveness. The first investments in this new technology were done without any empirical evidence of the expected benefits. In 1987, one of the largest U.S. trucking companies equipped 5000 vehicles with positioning and communications devices. Shortly after that, the U.S. Department of Energy (DOE) introduced the Transcom system for tracking radioactive materials on rail cars and trucks.

Since then, benefits such as cost reductions and efficiency improvements have been reported for the pre-shipment, in-transit, and post-shipment stages, as summarized in Figure I.12 [Batz, 1991], [Morlok and Halowell, 1989]. One trucking company even reported even that the use of satellite tracking systems helps meet strict federal reporting requirements for hazardous loads. Complying with different regulations is a major issue when traveling long distances. Local pick-up and delivery business is also expected to become more flexible by employing advanced technology.

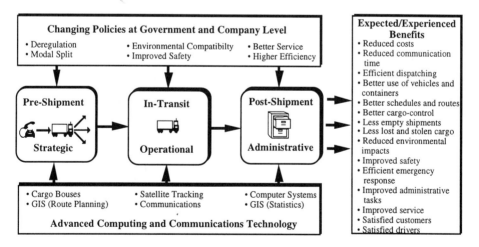

Figure I.12: The use of advanced communications and computing technology to benefit from changing policies.

Operational Risk Management

These reported benefits, however, consist of isolated experiences and do not reflect a representative picture of the transportation industry. For example, one transportation company reports that for a fleet of 640 vehicles the technology would be too expensive. Nevertheless, the reported benefits could motivate the transportation industry to adopt advanced technology as a standard, forcing shippers to employ it regardless of economic, environmental, safety, or other aspects. Further practical and theoretical studies need to be done before conclusive results concerning the benefits of using advanced technology can be presented. The central issues are being addressed in various pilot studies, such as DRIVE, Prometheus, Eureka, Carminat, and Europolis [Keen, 1992]. The objective of the DRIVE project is to come up with a European policy for using advanced telematics in the transportation industry.

In the late 1980s, the Joint Research Center, Commission of the European Community, Ispra, Italy developed a prototype risk management system called IRIMS (Ispra Risk Management Support System). It contains several simulation models which can be used for environmental assessment, risk analysis and system optimization. The system was implemented on a SUN-3/160 work station running under the UNIX operating system. Its architecture consisted of several modules, such as a hazardous material database, an industrial accident report, a chemical industry database, a transportation risk/cost analysis (including optimal routing), and an environmental impact assessment. One goal of the project was to improve the system by coding the mapping capabilities of an advanced GIS [Peckman et al., 1988].

As previously noted, the Association of American Railroads and the Railway Association of Canada are jointly developing a real-time, satellite based Advanced Train Control System (ATCS) for safer and more productive rail operations. The ATCS consists of (1) a color graphic display and data radio, monitoring the health condition of the crew, (2) a multisensory position information system, including transponder or satellite, (3) a two-way digital communications network for the rail system, (4) wayside switch and detector interfaces, and (5) a central computer. Operations were made more productive by (1) sophisticated traffic planning and supervision, (2) new train information for the dispatcher and train crew, including a dynamic track map and locomotive status, and (3) car pick up instructions channeled from the railroad order desk directly to the locomotive cab [Burns, 1989].

Currently, GPS technology is being used to improve safety through concepts as positive train separation (PTS) and positive train control (PTC). PTS computes safe breaking distances in real-time using dynamic and static information about train movements and the track system. PTC extends PTS in that it provides a dispatcher with information that allow the optimization of schedules [Phillips, 1998].

The University of Calgary developed a land based system to assist automobile drivers in finding the optimal route, given the current position and a destination. The system allowed a user to (1) position a vehicle using signals from satellites and information from on-board differential positioning devices, (2) plot the position on a flat panel display, (3) call up a digitized-electronic map of the area and see the vehicle's position relative to a desired location, and (4) obtain instructions (visual or audio) using an expert system on how to proceed in an optimal manner from the present to the desired location [Harris et al., 1987].

A potential application of GIS with reasoning capability is in emergency management. Several fire departments and police centers have stored data for emergency response into a GIS environment. Such a "graphical database" has a wide variety of applications. One important convenience of these mapping systems is their graphical display capability. The user can zoom into detail, move the map in all directions, magnify selected regions, add or remove specific data sets, etc. This capability of graphical scanning is a critical step in risk assessment for transit control.

The AURA project in Spain (Automatic Regulated Urban Access) is an example of the implementation of knowledge-based technology with weather data for real-time traffic control. Sensors were installed along two highways to gather real-time data on road surfaces (dry, wet, frozen), meteorological conditions (visibility, temperature, humidity), and traffic pattern (intensity, occupancy, speed, structure). These data are transmitted through a network to a processing center where recommendations are sent back to signal screens along the highway. These recommendations were aimed at informing the driver about opening and closing of new lanes, recommended speed, and head light signals for gaining access to the freeway. The system should be able to "reason" about anticipated traffic problems, such as traffic density and accidents, and provide appropriate guidance to the driver [Cuena, 1988].

A powerful and effective method of acquiring data in real-time is aerial reconnaissance and photo interpretation which originated during the second world war. By 1966, satellite imagery expanded the capabilities of aerial data acquisition even further. There is no doubt that this technology would be very useful for hazardous material flow control [Brugioni, 1988].

Other sources for data acquisition (in real-time) are the police; the vehicle/vessel itself, which can transmit observations to the central control unit through an on-line connection; and any other observers who might be concerned with hazardous material transportation. The control center will have databases on such characteristics as the environment surrounding various transportation routes, endangered environmental regions, ground water, and addresses of contacts in cases of emergency.

Changing national transportation policies and deregulation will promote long-distance shipments and enhance the local pick-up and delivery business. Shipping companies must also define new policies in order to survive in the highly competitive transportation industry. Third parties are pushing national governments and the transportation industry to comply with stringent environmental and safety requirements.

Employing advanced communications and computing technology seems to provide an answer that could satisfy the owners of the cargo, the shippers, and the third parties. However, four major research issues must first be addressed.

The first issue is the identification, development, and assessment of approaches for coupling intelligent decision support methodologies with advanced technology. These approaches will eventually be implemented into intelligent operational management systems that relieve the dispatcher as much as possible of his or her daily tasks. Although automation is a justifiable objective, the dispatcher will always be an integral part of the transport operation. Intelligent decision support must therefore be tailored to the dispatcher-computer system, providing optimal task sharing between man and computer. This novel dispatching environment will shift the emphasis from strategic planning to operational management. However, the appropriate design and the potential benefits of human-machine systems must be assessed prior to their implementation in an experimental setting or their testing by simulation.

The second issue is the definition of a policy for promoting the appropriate use and preventing the potential misuse of advanced technology. The employment of advanced communications and computing technology in the transportation industry will lead to new standards. These standards are now being set solely by the economic needs of shippers; however, safety, environmental, and other aspects must also be part of a comprehensive approach. This implies, for example, that operational reroutings must be done at the cost of the carriers, if the situation requires it. To enforce such a practice, an international policy must be defined and enforced. This will require a thorough analysis of the potential benefits and downfalls of using advanced technology in transport operations.

The third issue is the identification and assessment of strategies for development and implementation of standards for electronic data interchange. In order to gain maximum benefit from emerging information and communications technology, an effective coordination among independent agencies and specialists must be established. This coordination is necessary to facilitate or even to create the possibility for data exchange in the transportation business.

The fourth issue, finally, is the design of the transition phase for implementing intelligent operational management systems. Today's dispatching operations use little or no advanced technology, and decisions are based on the dispatcher's intuition and experience. Introducing advanced technology and novel decision support approaches requires careful consideration of the transition process from traditional to new processes. If the dispatcher does not learn to appreciate the benefits of the new approach, even the most sophisticated system risks being left idle.

CHAPTER II: MATHEMATICAL EXPOSITION OF OPERATIONAL RISK MANAGEMENT

1. The Graph Theoretic Approach of ORM

1.1 The Topological Graph Structure

Reasoning in operational risk management consists of assessing the consequences of **real-time events** (RTEs) on **attributes** (e.g., risks and costs), evaluating alternate courses of action, and deciding either to remain on the present or to take a new course of action. Analysis and decision making must be done in real-time. It is presumed that a human operator cannot both assess effects of RTEs and generate and evaluate alternate courses of action without support. Information technology can perform computationally intensive tasks in support of human information processing and reasoning. In order to successfully blend human and machine capabilities, a decision structure must be designed.

Sage [1986] notes that operational decisions are more likely to be made within a set of guidelines than strategic planning decisions. Such a meta-reasoning structure must allow one to assess the impacts and re-evaluate a given course of action whenever an RTE occurs. It also must consider that humans in a real-time decision situation must perform under conditions of uncertainty and time constraints. A graph theoretical approach will be used to structure real-time risk analysis and decision making situations.

A **course of action** in operational risk management consists of a temporal ordered sequence of decisions and concomitant selected activities. Each activity is preceded by a decision, which in turn leads to a new decision on the next activity. Examples of activities in the case of hazardous materials transportation are driving along different road sections, loading a vehicle, and refueling the vehicle. Examples of activities in the case of emergency response include activating the response team, closing the harbor, and using dispersants. Each activity is preceded by a decision, that is, every specific activity has been chosen from a set of several possible activities. Some activities require that other activities be completed; others do not.

The first decision to be made in operational risk management is to decide whether a significant event has occurred and, if so, to reevaluate the present planned course of action. The last decision is to decide if "normal" operation can be resumed, that is, whether the hazard has been resolved. It is important to note that there are always several different ways to abate the risk for a particular RTE. In other words, there are different courses of actions that could be taken.

Courses of action can be represented on a graph structure. A graph consists of nodes (vertices) and links (edges). Nodes represent the decisions and links the activities. An oriented link from decision node d_i to decision node d_j indicates that decision d_i has been made to do activity a_{ij} which leads to decision d_j; that is, the link a_{ij} states that there is a relation between the two decisions d_i and d_j. Therefore, every activity leads to a subsequent decision, except for the last decision stating that normal operation can be resumed (see Figure II.1).

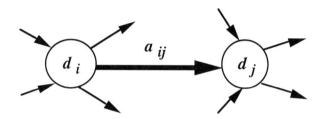

Figure II.1: The topological graph structure.

The graph with oriented links will be called the **topological graph structure** of real-time decision making. A feasible course of action is a sequence of activities on the graph which begins at the first decision node (i.e., start of operation), continues along intermediate activity nodes (considering the orientation of the links), and ends at one of the end stages (i.e., end of operation). Selecting an activity a_{ij} can also be seen as a walk on the graph from decision node d_i to decision node d_j. A course of action is therefore a walk through the graph from origin (first node) to one of the destinations (last node). Depending on the graph, there might be no, one, or many such walks. The determination of all feasible walks (courses of action) is a combinatorial problem on the topological graph. The human operator, considering all RTEs, specifies all feasible walks through the graph (courses of action).

The feasibility of a link (activity) can depend upon external conditions or preceding courses of action. Using a transportation network as an example of a topological graph structure with the nodes being cities and the links connections between two cities, note that some roads require special permits for shipping hazardous materials. Thus, the feasibility of using such roads depends on what cargo is being shipped. In emergency response, the feasibility of links (activities) is usually dependent upon previously selected activities. For example, the decision to use chemicals to extinguish a fire can only be made if the appropriate equipment is available.

Let $D=\{d_1,...d_n\}$ be the set of all possible decisions in operational risk management, and $\Re$ a binary relation on D, $(\Re,D)$: $d_s\Re d_t=a_{st}$, stating that decision d_s leads to decision d_t by taking activity a_{st} (see Figure II.2).

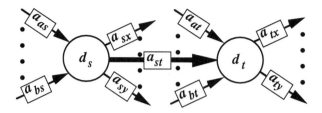

Figure III.2: Course of action.

The total space of possible CAs consists of a finite non-empty set $D=\{d_i\}$ of decisions together with a relation $\Re$ on D. The relation $\Re$ has the following characteristics:

- it can be irreflexive ($\neg\, d_i\Re d_i$)
- it is always transitive $(d_i\Re d_j \wedge d_j\Re d_k \rightarrow d_i\Re d_k)$
- it is not necessarily symmetric $(d_i\Re d_j \rightarrow d_j\Re d_i)$

Every relation $d_i\Re d_j$ stands for an activity a_{ij} as part of the operational measure. The activity a_{ij} is said to be **incident from** d_i and **incident to** d_j. Two subsets of D are fundamental: $S \in D$ is the set of decisions to start operational risk management and to begin with a strategically planned procedure (i.e., an RTE has been perceived), and $E \in D$ is the set of decisions to end operational risk management (e.g., a hazardous material shipment has reached its destination, or an emergency response has been successfully completed).

A CA consists of an ordered sequence of decisions and concomitant selected activities, e.g., CA=$d_1,a_1,d_2,a_2...,a_{n-1},d_n$, beginning with the decision $d_1 \in S$, stating that an RTE has occurred, and ending at the decision $d_n \in E$, stating that the objectives of the operation have been met, such that for $1 \leq j < n$, a_j is incident from d_j and incident to d_{j+1}. Each activity in operational risk management is, therefore, proceeded by a decision to take this activity and leads to a new decision regarding the following activity. Therefore, decision d_i has been made to undertake activity a_{ij} which leads itself to decision d_j.

A **sub-course of action** (SCA) is an ordered sequence of decisions and concomitant selected activities that does not necessarily contain elements out of S and E. A course of action, CA$^{(k)}$, with k decisions and k-1 activities is said to be of order k (k-order CA). In terms of the decisions, it can be written in the following way: CA=$\{s_1,...,d_i,...,e_k\}$. In terms of the activities, it can be written as

$CA = \{a_1,...,a_j,...,a_{k-1}\}$. The last decision, e_k, does not lead to a new activity but rather to the end of the operation at a satisfactory operational level.

Two CAs are **decision-disjoint** if they do not have any decisions in common. Two CAs are **activity-disjoint** if they do not have any activities in common. An operational risk management problem is called **connected** if there is at least one SCA between any two decisions.

Theorem 1: Two decision-disjoint courses of action are activity-disjoint.

Proof: Case 1: a_{ij} is a common activity of CA_1 and CA_2. Let it be incident from d_i^1 and incident to d_j^1 of CA_1, and incident from d_i^2 and incident to d_j^2 of CA_2. It then follows that at least d_i is common. Case 2: d_i is the only joint decision point (which is incident to different activities). In this case, no activity can be common. For example, shipping explosives to some harbor on a tanker (as part of CA_1) and fertilizers on a small boat to the same harbor (as part of CA_2) leads to the same decision point (whether to unload to train or truck). However, if this is the only common decision point, it means that explosives and fertilizers cannot continue the journey on the same mode.

An important class of connected operational risk management problems is the one where all possible SCAs with the same start-decision lead to different end-decisions. These problems will be called **prestructured** ORM problems. In graph-theoretical terms this family of ORM is represented by trees.

Theorem 2: In a prestructured operational risk management (ORM) problem, there is only one SCA starting at some decision d_i and leading to one of the ends of operational risk management.

Proof: Let SCA_1 and SCA_2 be two different SCAs, starting at some decision d_i and leading to the same end-decision d_e. In order to be different, they need to merge somewhere in d_j (latest in d_e). The set of decisions from d_i of SCA_1 to d_j and the set of decisions from d_i of SCA_2 to d_j form two new SCAs with common end-decision. This, however, is not possible.

1.2 Solvability of Operations

An important characteristic of an ORM problem is the number of feasible CAs. If the number of feasible CAs is small, then the ORM problem is said to be

sensitive to RTEs; if the number is large, the problem is said to be **stable**. In order to define a measure of stability for ORM, the **activity matrix** A is introduced. The element $a_{ij} \in A$ is 1 if $d_i \Re d_j$, and 0 otherwise; that is, if the decision d_i can be made to take activity a_{ij} that leads to decision d_j. Let

$$a_{ij}^k = \sum_{r=1}^{n} a_{ir}^{k-1} \times a_{rj}$$

be the k-th matrix product of the activity matrix A, where $a_{ij}^0 = 1$, and $a_{ij}^1 = a_{ij}$.

Theorem 3: The element a_{ij}^k is the number of k-order SCAs from d_i to d_j. This is a well known result from graph theory [Gibbons, 1989].

Proof: By definition, if $k=1$, then $a_{ij}=1$ if $d_i \Re d_j$, and 0 otherwise. Let's assume that the theorem holds for all powers smaller than k. Therefore, a_{ir}^{k-1} is the number of k-order SCAs from d_i to d_r, and $a_{ir}^{k-1} \times a_{rj}$ the number of k-order SCAs with a_{rj} as last activity. The summation over all decisions incident to d_j gives the result.

An example of an emergency response space for the cleanup of an oil spill is given in Figure III.3.

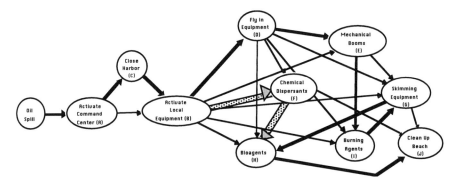

Figure III.3: Emergency response space.

The number of k-order CAs is called **solvability**, $s^{(k)}$. If $s^{(k)}=0$, then there is no way to solve the operation from the start point (oil spill) to the end point (clean up beach). The number of k-order SCAs and the solvability $s^{(k)}$ (number of k-order CAs from the origin (oil spill) to the end of the emergency response (clean up beach) are given in Table II.1. For example, among the 11 decision nodes there are 23 actions, that is, activities of order $k=1$. Moreover, there are 78 SCAs of order k=4, and $s^{(4)}=3$, where a CA starts with an oil spill (start decision) and ends with the clean up of the beach (end decision).

k	1	2	3	4	5	6	7	8	9
$\sum_{j=1}^{11}\sum_{i=1}^{11} a_{ij}^k$	23	43	65	78	71	48	23	7	1
$s^{(k)} \equiv a_{start,end}^k$	0	0	0	3	10	14	11	5	1

Table II.1: Number of k-order SCAs and CAs.

The **solvability index** $p(k)$ is defined as the probability of having a CA consisting of k actions. For an activity space consisting of n activities, we have:

$$p^{(k)} = \frac{s^{(k)}}{\sum_{k=1}^{n} s^{(k)}}.$$

For the activity space of Figure III.3, we have $\max(p^{(k)})=p^{(6)}=14/44=0.32$.

Both for the strategic planning of CAs and for the revision of CAs when RTEs occur, one is interested not only in the feasible CAs but also in the optimal CA out of all feasible CAs. In order to compute the optimal CA, a preference structure must first be overlaid on the activities of the ORM problem. The preference of activities, as well as the adjacency of decisions (i.e., the existence of activities) can be conditional or unconditional. Conditional preferences can depend on previous activities and decisions or on state variables, such as weather and environmental conditions. A course of action is said to be affected by an RTE, if the preference or existence of at least one activity is affected by the RTE. In the following section, a preference structure will be introduced to assess preferences of activities and revise them given an RTE. Thereafter, a procedure for reassessing preferences will be given and an algorithm to compute optimal CAs, both for conditional and unconditional CAs.

1.3 Preference Classes

Every activity has preferences that can be expressed in terms of a set of attributes. These attributes are given values by the decision maker. In operational risk management, the main attributes are risk, costs, and benefits. Risks refer to human life, environmental pollution and destruction, property loss (also expressed in costs), loss of company image, etc. Costs are expenditures for a given activity, while benefits are the positive impact of an activity, such as the amount of barrels of oil removed from the shore. It is assumed that both costs and benefits can be estimated with sufficient accuracy to be considered deterministic. Risks during operations are typically characterized by high uncertainty and imprecision, especially when evaluated under time constraints and stress. Although operational risks are non-deterministic, no explicit stochastic model can be postulated. Therefore, emphasis is placed on the possibility of real-time events and their impacts rather than on the events' expectation in a probabilistic sense [Ostroff, 1989].

For each operation that is affected by an RTE, alternative CAs must be evaluated, from the current status to the planned ending of the operation. The set of "best" CAs (which is a subset of the set of feasible CAs) can then be determined. The expected results of activities are measured by specifying an ordered set of preference classes for the set of attributes. Ordinal measurement means that each attribute must be assigned to one and only one class and that these classes can be ordered. Risk is incommensurable among classes, that is, there exists no number n that can measure the risk among different classes, but risk could be measured with a cardinal scale within a class. Therefore, it is assumed that it is only possible to assess whether one activity is more or less dangerous than another one.

Ordinal relations will be used not only for the assessment of operational risks but also for the tradeoffs between risks and any other attributes, such as costs, benefits, or values. In this section, the procedure of defining the preference classes and their ordinal relations is described for the case where only two attributes, "cost" and "risk," have been identified.

As an example of determining the preference tradeoffs, let's consider the case of hazardous materials transportation. Both the risks and costs for shipping these materials have been discussed in the literature. One assessment is that the average accident probability is 10^{-6} per truck-mile, and the risks lie between 10^{-8} and 10^{-10} deaths per truck-mile [Glickman, 1989]. Therefore, single shipments of even the most hazardous materials (poison gas and explosives) are considered to be "low risk" even when the route goes through highly populated areas [Russell et al., 1986].

Despite this "low risk" level, the potential for catastrophic events is inherent in every shipment. Risks that are very low and highly uncertain and have the potential for catastrophic consequences are difficult to assess numerically. Apostolakis [1978] notes that numerical risk estimations for values below 10^{-4}

are suspicious because of the high uncertainty. In addition to objective risk assessment, Urbanek and Barber [1980] suggest that subjective factors be also considered. In the case of hazardous material transportation, subjective factors refer to special population (e.g., schools, hospitals, senior citizen homes), sensitive roadway structures which can increase the damages (e.g., bridges, tunnels, intersections, etc.), the location of the response units, and the carrier's time and travel costs. Therefore, it is assumed that in a real-time environment the operational risk manager can only assess whether an RTE causes an action (e.g., which is safe under normal conditions) to become low-risk (LR) or high-risk (HR), where low risks are in the order of magnitude of 10^{-10} per vehicle-mile, while the order of magnitude of high risks is assumed to be 10^{-4} per vehicle-mile.

In order to define a tradeoff between risks and costs, life-saving costs must be considered. Life-saving costs are the costs society is willing to pay in order to save one (probabilistic) life. These costs depend strongly on the nature of the risk. Fischhoff et al. [1981] indicate that U.S. society spends about \$140,000 in highway construction to save one life but \$5 million to save a person from death due to radiation exposure. Considering the high public aversion for hazardous materials accidents, it is assumed that life-saving costs for hazardous materials transportation are in the order of magnitude of 10^7. This implies that 10^{-3} $(=10^{-10} \times \$10^7)$ per mile should be spent to avoid LR actions and 10^3 $(=10^{-4} \times \$10^7)$ per mile to avoid HR actions.

Because the transportation costs are about \$1.00 per mile [Abkowitz and Cheng, 1988] and the average trip length less than 500 miles, we expect that the cost of avoiding a specific activity is less than \$500. Consequently, a HR action must whenever possible be avoided regardless of the costs (if there is a feasible route on the transportation network; i.e., a feasible SCA), and a LR action will be avoided only if doing so would incur no cost increase. From a descriptive point of view, we would therefore look for a CA by minimizing the number of HR actions, then break ties by minimizing costs, and, if there are still ties, minimizing the number of LR actions.

Because the ordinal approach to preference assessment of actions is based on perception and subjective factors, it is more appropriate to talk about the preferences of actions with regard to risk and cost aspects. Therefore, a safe and low-cost route has a high preference and an unsafe and high-cost route a low preference. The overall preference of a hazardous materials shipment route is the "sum" of the preferences of all the entities along the route.

For this discussion, three preference classes will be delineated: High-Risk (HR), Low-Risk (LR), and transportation Costs (C). Two more classes will be introduced: the α preference stands for actions that are either too hazardous or too expensive and therefore must be avoided under any circumstances; that is, if there is no feasible route on the transportation network that avoids entities of preference α, the shipment must stop. The preference ω stands for actions that are of no safety or cost concerns; that is, the entities that were initially identified as potential concerns but then later perceived to be neither hazardous nor costly. Therefore, the **preference classes** have the following ordinal relations, where '$\prec$' means 'less preferred than:'

$$\alpha \prec HR \prec C \prec LR \prec \omega.$$

Past research has showed that in stress situations (such as during RTEs), serious decisional conflicts arise if decision makers become aware of the risks and that they tend to make choices according to outstanding characteristics [Ben Zur and Breznitz, 1981]. Therefore, risk/cost assessment during RTEs must be pre-structured, by asking the dispatcher to assign every entity to one out of four preference classes for risk and one out of three preference classes for costs. The set of risk preference classes to choose from is $\{\alpha, HR, LR, \omega\}$. The set of cost preference classes to choose from is $\{\alpha, C, \omega\}$, where C is a monetary unit, for example, $1.00. Thus, a pair of risk/cost preferences is assigned to each identified activity. While a multiple assignment of C is allowed (e.g., $20.00) for cost preferences, this is not possible for risk-preferences.

For operations other than hazardous materials transportation (e.g., emergency response) the same approach is applicable. It is assumed that the operational risk manager is faced with activities that must be considered feasible, but have risks so great that there is no monetary equivalent. However, there are also risks that do not justify the expenditures needed to abate or avoid them. Activities that do not have any significant risks or costs will be assigned to the class ω; activities that may not be engaged in under any circumstances will be assigned to the second class α.

Note that some activities, however, will have a positive impact, a benefit. Therefore, a benefit class can be added to the preference structure. A possible preference structure for the case of emergency response is the following:

$$\alpha \prec \text{High-Risks} \prec \text{Benefits} \prec \text{Costs} \prec \text{Low-Risks} \prec \omega.$$

Since an activity with low risks is preferred to an activity with high risks, the ordinal relation implies a lexicographic preference structure. Given the topological graph structure, the operational risk manager must map the attributes of all activities into the preference classes. Just as with the feasibility of activities (links), the preference for an activity can depend on previous courses of action.

For example, using chemical dispersants to respond to an oil spill has a smaller risk (higher safety-preference) if the effectiveness of the dispersants has been previously tested. After all activities have been classified using such a preference structure, the graph is called a **preference graph**.

Whenever an RTE occurs, the operational risk manager must reassess risks, costs, and benefits of all activities affected by the RTE using the defined preference classes – in real-time. This amounts to reassessing the preferences of the affected activities considering the impact of the RTE. In order to reduce the complexity of the reassessment process, decision support can be provided by having a computer perform the calculations and search for new CAs. The operator can determine the affected links (activities) by noting them on a graphical display. If the identified actions affect other actions, both the feasibility and preference of these actions must also be reassessed. If, for example, the RTE is "an icy road" then the preference of using a truck to disperse chemicals for fire-fighting will be lower because of the possibility of an accident. Therefore, the preference of using chemicals to fight the fire will also be lower. All the affected activities are presented to the operator one-by-one or grouped into similar operations for reassessment of their preferences. Because the computer does the tedious error-prone search of activities that have to be reassessed, the operational risk manager can concentrate fully on the assessment itself. This process of real-time risk analysis is supported by the use of interactive multimedia displays.

2. Individual Decision Maker Situation

2.1 Cognitive Assumptions for Preference Assessment

An operational risk manager who is forced to make decisions in real-time that may involve loss of life or property, or have a disastrous impact on the environment, can not be expected to be rational. In fact, research has shown that operational risk managers select alternative courses of action based on very few characteristics or even randomly [Ben Zur and Breznitz, 1981], [Janis and Mann 1977], [Belardo et al. 1984].

One of the more disturbing conclusions of recent review of the research on assessment and choice for decision making was that no simple, general model exists for describing how preferences of individuals are determined [Kleindorfer et al., 1993, p. 175]. In fact, Einhorn and Hogarth [1981, p. 61] noted in their review of behavioral decision theory that "the most important empirical results in the period under review have shown sensitivity of judgment and choice to seemingly minor changes in tasks." Therefore, each task has its own unique set of characteristics and associated organizational considerations and constraints. As noted in the previous chapter, operational risk management deals with sudden onset, very low probability, high consequence events which must be responded to in a timely, accurate manner.

De Keyser [1987] states that operators in an emergency situation rarely reason in a deductive way but rather act as furious pattern matchers. Wagenaar and Groeneweg [1987] show that people construct a hypothesis by looking only at the most salient symptoms, and thereby forget contrary evidence. Moray [1987] postulates that the operator has a mental model of the system that consists of a set of quasi-independent subsystems that do not correspond a to a one-to-one mapping of the real system. Mancini [1987] concludes that decision aids for crisis management have to take into account the cognitive aspects of human behavior. This is done by first structuring the cognitive assessment and decision making process and then postulating cognitive assumptions about the human operator.

Operational decision making is complex in general, due to the high number of feasible CAs and their implicit representations. CAs may not be explicit, and may have to be represented in a form helpful to the operational risk manager. For example, an emergency response manager may forget that mechanical booms are required to remove oil on the water before it can be burned. The graph representation defining the relations among the various activities that could comprise a CA must be constructed either a priori as part of the strategic planning, or during the response to the RTE in constructing new activities that could result in new CAs.

The evaluation of CAs can be done in two ways: (1) by assessing the impacts of the RTE on the attributes of the affected activities and then computing new optimal CAs with an appropriate algorithm: **evaluation by attribute**; or (2) by constructing a few feasible CAs (e.g., using a simple heuristic) and comparing them to each other: **evaluation by alternative**.

When evaluation by attribute is completed, the operational risk manager could decide not to recommend the new optimal CAs. One reason might be that the operational risk manager wants to continue to follow a certain CA, although the choice model suggests that a new CA be selected. In such a case, s/he might change the preference of some attributes of certain activities that seem not to be assessed properly, given the new CA that was determined by the decision model using the initial assessments. This iterative decision making approach will be repeated until the new CAs are satisfactory to the operational risk manager or until the decision is made to evaluate CAs by alternative (Figure II.4).

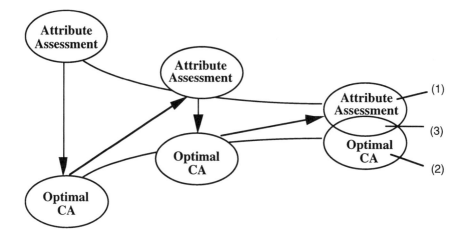

Figure II.4: Iterative decision making approach for evaluation by attribute.

There are three different possibilities for evaluation by attribute before the operator recommends a new CA: (1) the new CA determined by the decision model is not creditable to the operator and s/he chooses a different one (suboptimal behavior); (2) the iterative reassessment of the attributes produced a satisfactory solution [Simon, 1972], but the assessment itself does not seem to be appropriate to the operational risk manager anymore (suboptimal behavior); or (3) the assessment by the operator and the determined CA by the decision model both satisfy the operational risk manager (optimal behavior).

The decision situation faced by the operational risk manager is characterized by task complexity due to time pressure and the large number of alternative CAs. In such a setting, two decision strategies seem to be used: (1) attribute-based assessment instead of assessment by alternative, and (2) non-compensatory strategies instead of compensatory. Early research by Payne [1976] found that attribute-based processing had shifted to non-compensatory strategies when he increased the number of alternatives. Later work by Payne and Braunstein [1978] supported these results. Research by Zakay [1985] and Payne et al. [1988] found similar results under conditions of time pressure. Additional research on both these topics can be found in [Payne et al., 1993]. As will be discussed in Chapter VII, similar results were found in an experimental evaluation. These results are incorporated into the formalization of operational risk management by stating basic assumptions about how an operational risk manager assesses and reasons. These cognitive assumptions define the assessment process and the reasoning logic for choice in an operational setting.

Cognitive Assumption 1: *Evaluation*. It is assumed that in operational risk management situations evaluation by attribute is more accurate and requires less effort than evaluation by alternative. This assumption is tested in the experimental assessment in Chapter VII. Therefore, the decision variables used in selecting an optimal course of action are the attributes. The activities are assessed by evaluating their attributes.

Cognitive Assumption 2: *Measures for Assessment*. It is assumed that experienced operators prefer to assess risks (attribute with high uncertainty) in terms of well-defined ordinal preference classes rather than relating them to a numerical scale. Costs (attribute with low uncertainty) are assessed on cardinal scales. The ordering function $\phi(a_{ij}|H): a_{ij} \rightarrow \Pi^{(k)}$ assigns an activity a_{ij} to a preference class $\Pi^{(k)}$, given a **history** H. The assessment spectrum Π can therefore be defined as an ordered set using the relation $\Pi^{(k)} \prec \Pi^{(l)}$, which means "$\Pi^{(k)}$ is less preferred than $\Pi^{(l)}$": $\Pi = \{\Pi^{(1)} \prec ... \prec \Pi^{(n)}\}$. The class $\Pi^{(k)}$ is said to be "lower" than the class $\Pi^{(l)}$. These classes represent for attributes, such as Cost, High-Risk, and Low-Risk. This assumption has also been evaluated in an experimental setting; the results are presented in Chapter VII.

The history of preferences, H, reflects all the conditions under which this preference holds. These can be activities previously taken (**operational history**) or environmental conditions, such as weather, time, etc. (**state history**). Time dependency is crucial in operational risk management of large scale systems. It is characterized by two aspects: the location of the operation in space and the actual status of the operation. The preference for an activity as a function of its geographical location fluctuates in time. For example, the activity "driving by a school" may be safe at night (high preference) but not during the day (low preference). Because the operational risk manager may not know the time of day the activity takes place, the preference must be expressed dynamically, even for the "normal" situation when no RTEs are present. This is called the **spatial aspect of time**.

In addition, the preference for an activity depends on its status. The preference for driving by a school during the day depends also on the number of students in the school and the day of week. In the "normal" case, this time dependency is taken into account in planning a route. But it might also be the situation where several RTEs already affect an activity. If an additional RTE affects this activity, the operational risk manager has to consider the whole situation including earlier RTEs. This is called the **contextual aspect of time**. Therefore, if RTE_o is an ongoing real-time event and RTE_n is a new real-time event that forces the operational risk manager to assess the affected activities, the conditional preference assessment function for activity a_{ij} is $\phi(a_{ij}|RTE_o, RTE_n)$.

Cognitive Assumption 3: *Insignificance of Entities (De Minimis)*. An operational risk manager assigns a ω-preference to an activity, knowing that a CA consisting of any number of ω-preferences is the most desirable CA.

Cognitive Assumption 4: *Avoidance of Activities*. An operational risk manager assigns a α-preference to an activity, knowing that no CA can contain this activity; that is, the activity becomes unacceptable and any CA containing this activity is not feasible.

2.2 Preference Algebra and Reasoning Logic

The foregoing discussion presented the topological graph structure, the preference classes for operational risk management, and the cognitive assumptions for preference assessment by the operational risk manager. For example, in the case of hazardous materials transportation by truck, the graph structure is the road network, with the major activities "driving along various road segments."

The preference of a course of action with n activities, given the preferences of all activities, π_{ij}, is defined as: $\pi_{CA} = \pi_{12} \oplus ... \oplus \pi_{n-1\,n}$. In order to be able to compute optimal courses of action, or to compare two activities and "add up" the preferences of two activities, a preference algebra must be introduced.

The **preference algebra** is defined as a system $<\Pi, \prec, \oplus, \alpha, \omega>$. It consists of an infinite set of preferences $\pi_i \in \Pi$, the relations $\prec$ (preference) and $\sim$ (indifference) in the preference algebra, an operator $\oplus$ on couples of elements of Π, and two distinguished elements $\{\alpha, \omega\} \in \Pi$. The set Π is closed under the operator $\oplus$.

The preference α stands for activities that must be avoided and the preference ω stands for activities that are of no concern. The set Π consists of any number of compound preferences.

For $\pi_i, \pi_j, \pi_k \in \Pi$, the operator $\oplus$ is:

- monotone: $\pi_i \prec \pi_j, \pi_k \; \alpha \to \pi_i \oplus \pi_k \prec \pi_j \oplus \pi_k$; holds also for "$\sim$"
- commutative: $\pi_i \oplus \pi_j \sim \pi_j \oplus \pi_i$
- associative: $(\pi_i \oplus \pi_j) \oplus \pi_k \sim \pi_i \oplus (\pi_j \oplus \pi_k)$

The relation $\prec$ is:

- irreflexive ($\neg\ \pi_i \prec \pi_i$)
- transitive ($\pi_k \prec \pi_i \wedge \pi_i \prec \pi_j \rightarrow \pi_k \prec \pi_j$)
- asymmetric ($\pi_i \prec \pi_j \rightarrow \neg\ \pi_j \prec \pi_i$)

The relation $\sim$ is:

- reflexive ($\pi_i \sim \pi_i$)
- transitive ($\pi_k \sim \pi_j \wedge \pi_i \sim \pi_j \rightarrow \pi_k \sim \pi_j$)
- symmetric ($\pi_i \sim \pi_j \rightarrow \pi_j \sim \pi_i$)

Preference and indifference relations are complete ($\pi_i \prec \pi_j \vee \pi_j \prec \pi_i \vee \pi_j \sim \pi_i$).

This preference algebra allows the "addition" and the comparison of preferences of any two activities, SCAs, or any combination of them. To define the properties of ω, the following **axiom** is introduced:

$$\pi_i \oplus \omega \sim \pi_i;\ \pi_i \in \Pi.$$

Theorem 4: The preferred of two SCAs is the one with the fewest elements in the lowest preference class (for ties, the next higher preference class is used). This can be written as: $p\pi_i \oplus \pi_j \prec q\pi_i \oplus \pi_k;\ \pi_i, \pi_j, \pi_k \in \Pi;\ \pi_i \prec \pi_k;\ p, q \in N;\ p > q$.

<u>Proof</u>: It can be shown by induction that $k\omega \sim \omega;\ k \in N$. In addition, it can be shown that if $\pi_i \prec \pi_j$ and $\pi_m \prec \pi_n$, then $\pi_i \oplus \pi_m \prec \pi_j \oplus \pi_n$ and that $\pi_i \prec \pi_j \rightarrow n\pi_i \prec \pi_j$; $\pi_i \prec \omega,\ \pi_i, \pi_j \in \Pi,\ n \in \mathbb{N}0$.
From this it follows: $\pi_i \prec \pi_k \rightarrow (p-q)\pi_i \prec \pi_k \rightarrow p\pi_i \prec q\pi_i \oplus \pi_k$.
Case 1: $\pi_j \sim \omega \rightarrow p\pi_i \oplus \omega \sim p\pi_i \prec q\pi_i \oplus \pi_k$.
Case 2: $\pi_j \prec \omega \rightarrow p\pi_i \oplus \pi_j \prec q\pi_i \oplus \pi_k \oplus \omega \sim q\pi_i \oplus \pi_k$.

This theorem has several important implications. It first states that there is no number, n, of activities of a higher preference class that can be considered equivalent to one activity of a lower preference class. For the example of hazardous material transportation it means that high risk road segments must be avoided, regardless of the costs of rerouting. The theorem also allows the comparison of any two activities or SCAs. The principle of comparison corresponds to the lexicographic ordering or choice by first difference in the preference classes.

2.3 Finding and Revising Optimal Courses of Action

Operational risk management (ORM) consists of an **active** and a passive part, as shown in Figure II.5. It starts by activating a strategically planned CA. In the case of emergency response, the CA can be the evacuation of populated areas, the removal of oil on the sea, and the salvage of the vessel. In the case of transportation of hazardous materials, the CAs are the vehicles driving along the transportation network.

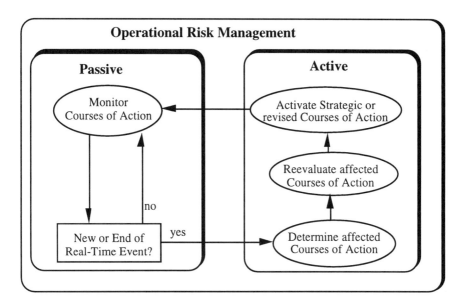

Figure II.5: Tasks and decisions in operational risk management.

As long as no RTEs occur, the operational risk manager monitors the execution of these CAs passively. If an RTE occurs, ORM becomes active. The operational risk manager must determine which CAs are affected by the RTE. In the case of hazardous material transportation the operational risk manager (dispatcher) must determine the vehicles that plan to drive through the area affected by the RTE (e.g., a snow storm).

A preference matrix Π is introduced which is similar to the activity matrix A introduced in Section 1.2. The element $\pi_{ij} \in \Pi$ is the preference for activity a_{ij}. The activity matrix can therefore be redefined in terms of the preference matrix:

the element $a_{ij} \in A$ is 1 if $\alpha \prec \pi_{ij}$, and 0 otherwise; that is, the activity a_{ij} is feasible if the preference is higher than α. By definition, α stands for activities that must be avoided; that is, activities that are not feasible.

As previously defined, a CA is affected by an RTE if the preference of at least one activity is affected. After the affected CAs, that is, the preferences of the affected activities, have been reevaluated, new optimal CAs must be determined. Then, these new CAs have to be implemented and ORM goes back to the passive stage.

Two procedures must be defined in this process: 1) how to determine the affected CAs; and 2) how to compute optimal CAs, given that all activities have been assigned multiple conditional preference values. The procedure for finding the affected CAs is outlined first. Every activity, a_{ij}, has assigned multiple preferences $\pi_{ij}^{(k)}$. These preferences can be conditional or not; that is, they can have a history consisting of activity and/or state dependency. Let $\pi_{ij}^{(k)} | (ah_{ij}^{(k)}, sh_{ij}^{(k)})$ be the k-th conditional preference of activity a_{ij}, where $ah_{ij}^{(k)}$ stands for the activity condition (i.e., the preference holds if the mentioned activities have been previously taken), and $sh_{ij}^{(k)}$ for the state conditions (i.e., the preference holds if the mentioned states, such as environmental conditions, time when the activity takes place, etc., hold). An RTE can affect state variables (e.g., weather) and/or activities. Every activity has a set of relevant state variables and relevant activities. If an activity or a state variable influences a preference for an activity, it is considered to be **relevant**. The **affected activities** are those which have at least one relevant activity or state variable affected by the RTE. The **affected SCAs** are those which have at least one affected activity.

The affected CAs must be reevaluated. This can be done, as previously mentioned, either by alternatives or by attributes. Based on Cognitive Assumption 1, the discussion is limited to the construction of CAs by attributes. This requires that the impacts of the RTEs on the attributes of the CAs be assessed first. This is usually done by revising the preferences for each affected activity one-by-one. After this reassessment, new optimal CAs must be computed.

The objective of the search problem for finding an optimal CA is to look for the CA with the highest overall preference from either the beginning of the operation or from the activity that is going on at the moment an RTE happened. The decision variables are the activities. Based on the preference algebra, a mathematical description can be given.

Let n be the number of decisions of the ORM problem, π_{ij} the preference of activity a_{ij}. The x_{ij}'s are the decision variables, which are 1 if activity a_{ij} is chosen for the most desirable CA and 0 otherwise. Then, the mathematical formulation of the CA problem is the following:

$$\text{maximize:} \quad \bigoplus_{j=1}^{n} \bigoplus_{i=1}^{n} x_{ij} \pi_{ij}$$

subject to:

- $\sum_{i=1}^{n} x_{ik} - \sum_{j=1}^{n} x_{kj} = \begin{cases} 1, & \text{for } k = \text{start} \\ 0, & \text{otherwise} \\ -1, & \text{for } k = \text{end} \end{cases}$
- $\alpha \prec x_{ij} \pi_{ij}$
- $x_{ij} \in \{1, 0\}$

The objective function (1) states that the optimal CA is the one with highest preference. The two summations over n are for the total ORM problem preference, where n is the number of decisions. Equations (2) - (4) define the space of feasible CAs. Equation (2) stands for connectivity of the CAs. Equation (3) states that no CA can contain an activity of a preference. Equation (4) states that the decision variables x_{ij} must be 1 if the activity a_{ij} is an element of the optimal CA or 0 if the activity a_{ij} is not chosen.

The algorithmic search for the optimal CA is different for ORM problems with conditional preferences and for those with unconditional preferences in the following ways.

Case 1: Unconditional Preferences

Let Π be the preference matrix for a specific ORM with n decisions, where π_{ij} is the preference of activity a_{ij}. Let $\pi_{ij}^{(k)} = \text{pref}\{\pi_{ij}^{(k-1)}, \pi_{ix}^{(k-1)} \oplus \pi_{xj}^{(k-1)}\}$, where $\pi_{xj}^{(0)} = \pi_{ij}$, and $\text{pref}\{\pi_r, \pi_s\} = \pi_r$ for $\pi_s \prec \pi_r$.

Theorem 5: The most desirable CA from $d_i \in S$ to $d_j \in E$ is $\pi_{ij}^{(n)}$.

Proof (by induction, as known from graph theory [Gibbons, 1989]). If $k=0$ then $\pi_{ij}^{(k)}$ is the preference for taking only activity a_{ij}. Assuming that the statement also holds for $\pi_{ij}^{(k-1)}$ then $\pi_{ij}^{(k)}$ is the higher preference for $\pi_{ij}^{(k-1)}$ and $\pi_{ix}^{(k-1)} \oplus \pi_{xj}^{(k-1)}$. According to the hypothesis of induction, $\pi_{ij}^{(k-1)}$ is the highest preference for the sub-course of action consisting only of decisions out of $\{d_i, ..., d_{k-1}\}$. However, if there is a more desirable path which also uses d_k as decision then its preference must be $\pi_{ir}^{(k-1)} \oplus \pi_{rj}^{(k-1)}$. After $\pi_{ij}^{(k)}$ has been constructed, every alternative is examined and the theorem holds. As it is well known from graph theory, this algorithm is of $O(n^3)$. A faster algorithm would be based on Dantzig's algorithm ($O(n^2)$) [Dantzig, 1975].

The approach with Dantzig's algorithm is the following. Assume that at some stage k in the computing process the preferences π_{OC_i} and the sequence of activities of the most desirable sub-course of action from the origin decision O to the decisions C_i's are known, where $C_i \in C=\{C_1,...,C_k\}$. The decisions in C are called closed decisions. Let $M_j \in M$ (set of adjacent decisions of C, $M_j \notin C$) with the preference $\pi_{C_iM_j}$ to take activity (C_i,M_j). The algorithm is the following.

(1) *Initialize*: Choose (close) the origin decision O.

(2) *Close Node*: Choose and close (the preference to the origin decision is definitive) as the $(k+1)$-th decision, M_j, where j satisfies $\alpha \prec \pi_{C_iM_j}$ and $\pi_{OC_i} \oplus \pi_{C_iM_j} \sim \text{Max}(\pi_{OC_i} \oplus \pi_{C_iM_r})$, where $i=1,2,...,j,...,k$, $M_r \in C$. Ties are broken arbitrarily. Restated, M_j is a direct successor of an already closed decision C_i. In addition, it has the highest preference to the origin decision among all the direct successors of all the closed decisions.

(3) *Loop*: If the destination decision D is not in C, then go to (2), i.e., if M_j D. The preference to go from the origin decision to M_j is $\pi_{OC_i} \oplus \pi_{C_iM_j}$ and the most desirable sub-course of action is via C_i.

Case 2: Conditional Preferences

In an ORM problem with conditional preferences, every activity has assigned a set of conditional preferences. Some of these preferences have been assessed after an RTE has occurred. The k-th conditional preference of activity a_{ij} is $\pi_{ij}^{(k)}|(ah_{ij}^{(k)}, sh_{ij}^{(k)})$, where $ah_{ij}^{(k)}$ stands for the activity conditions, and $sh_{ij}^{(k)}$ for the state conditions. Therefore, two activities can be concatenated only if they have the same history. In addition, it is assumed that an activity can be taken only once in a CA. Thus, the algorithm must first search exclusively for all feasible CAs and then choose the one with the highest overall preference as the most desirable one. Feasible CAs are constructed by exhaustively concatenating activities with common history.

A **predicate logic** notation uses the following two **predicates** to define a CA:

activity(decision,decision)
course_of_action(decision,decision)

The three **clauses** define an activity, a CA consisting of only one activity, and a CA consisting of multiple activities:

activity(d_x,d_y)
course_of_action(d_x,d_y) if activity(d_x,d_y)
course_of_action(d_x,d_y) if activity(d_x,d_z) and course_of_action(d_z,d_y)

To find all feasible courses of action that are initiated by the decision d_{start} and that lead to the satisfactory operational condition d_{end}, the goal needs to be introduced:

course_of_action(d_{start},d_{end})

The most desirable course of action among all the feasible courses of action is the one with the highest overall preference.

2.4 Relation to Multiattribute Utility Model

The ordinal preference model can be approximated by the following monotonically increasing multiattribute utility function, $u(\alpha,HR,C,LR,\omega)$, where higher utilities are preferred to lower ones. Let x_α be the number of α preferences, x_{HR} the number of HR preferences, x_C the number of C preferences, x_{LR} the number of LR preferences, and x_ω the number of ω preferences.

$$u(\alpha,HR,C,LR,\omega) = k_\alpha \times x_\alpha + k_{HR} \times x_{HR} + k_C \times x_C + k_{LR} \times x_{LR} + k_\omega \times x_\omega.$$

The scaling constants for the monotonically increasing utility function have the following values: $k_\alpha = -\infty$, $k_{HR} = -10^8$, $k_C = -10^4$, $k_{LR} = -1$, and $k_\omega = 0$. This model holds if a course of action has fewer than 10^4 activities from the same preference class. In such a case, the preference classes remain incommensurable. If one or more activities are of preference α, the utility becomes minus infinity; that is, the course of action is not feasible.

Risk managers often think in terms of a logarithmic scale. For example, an annual probability of death of 10^{-6} is considered to be reasonable, while 10^{-3} is high, and 10^{-8} is low. Using a scale from 0 (no risk) to 100 (basically certain death), any integer value $x \in [0,100]$ can be transferred to an absolute risk value $\lambda(x) \in [0,100]$ as follows:

$$\lambda(x) = \begin{cases} 10^{-[10-INT(x/10)]}, & \text{for } x \in Z = \{0,10,20,\ldots,70,80,100\} \\ \left[x - 10 \times INT\left(\dfrac{x}{10}\right)\right] \times 10^{-[10-INT(x/10)]}, & \text{for } x \in [0,100] \setminus Z^{\cdot} \end{cases}$$

$INT(x)$ means to take the integer part of a number; for example, $INT(3.8)=3$. Table II.2 shows some numerical examples for this transformation.

x	0	38	40	74	92	100
$\lambda(x)$	10^{-10}	8×10^{-7}	10^{-6}	4×10^{-3}	2×10^{-1}	1

Table II.2: Transformation of preferences into risk values.

These absolute risk values, $\lambda(x)$, could then be transformed into risk-cost values, if a tradeoff between risks and costs is defined. For example, assuming that $\$10^7$ should be invested to reduce the risk by 1 unit, then one possible monotonically decreasing multiattribute utility model (lower values are preferred to higher ones) would be the following:

$$u(c,\lambda) = c + 10^7 \times \lambda.$$

Chapter IV discusses such an approach where the tradeoff is based on life-saving costs.

2.5 The Process of Preference Assessment and Decision Making

The two tasks of an operational risk manager are the assessment of risk, costs, and benefits of actions; and the determination and choice of courses of action. The purpose of operational risk management is to support operators during RTEs which are perceived as serious changes in safety, operating costs, or benefits. Most of the time, however, the operational risk managers are merely monitoring the operations on a large screen, without active intervention. Figure II.6 illustrates the process of preference assessment and decision making.

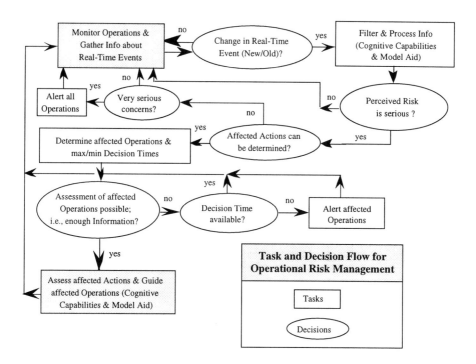

Figure II.6: Process of preference assessment and decision making.

Given the continuous update of real-time data and information, the tasks and the decision processes related to monitoring, assessment, and choice of courses of action must be analyzed from a dynamic point of view. The operational risk manager switches from the monitoring task to the assessment task only if an RTE seems to affect any of the operations with respect to any of the attributes. The occurrence of RTEs is unpredictable. Data and information about the RTEs can be very limited both in quantity and quality. The first task is therefore to filter and process any incoming data. Depending on the results, the operational risk manager might decide to immediately alert all operations or only those that may be affected by the RTE. An operation is affected by an RTE if any of the actions which are part of the planned CA are affected by the RTE.

An important parameter is the time it takes for the operations to be affected; that is the time left before the first affected action will be employed. If the operations won't be affected for a long time, the operational risk manager can postpone his/her decisions, giving priority to those actions which are already or soon will be affected by the RTE.

After the attributes of the affected actions have been assessed, new optimal courses of action must be determined for the affected operations. Three types of recommendations are possible. First certain operations can wait until the RTE is over; second, the planned CA can be changed to avoid the RTE; and third, the CA for some operations can progress as planned if the RTE is not too dangerous, costly, or benefit-negative. The suggestions for changing the CAs are then transferred to the operations managers (e.g., drivers of the trucks or on-site emergency managers).

3. Multiple Expert Decision Making Situation
3.1 The Decision Process for Multiple Experts

Operational risk management frequently involves multiple experts working at different locations. With the advent of technologies like mobile communications, experts can download data in the real-time, discuss ramifications for operations, and assess the current and proposed state of operations. Because different experts process incoming data differently, the results of their analyses and the recommendations for changing CAs cannot be expected to be the same. Discrepancies among the experts can be determined at two stages: (1) during analysis (i.e., assessment) and (2) when making recommendations for new or modified CAs (i.e., choice).

The experts are assumed to work independently but their results might not be of equal importance; that is, the assessments or choices of multiple experts might be weighted differently. The ways assessments and choices by n experts can be processed is illustrated in Figure II.7.

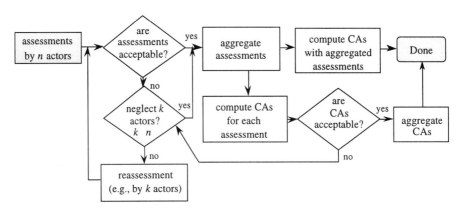

Figure II.7: Process of experts' assessment and choice.

The assessments by n experts are first checked for acceptability (which will be defined in the section to follow) even if they are not the same. If an assessment is determined not to be acceptable, the expert can either be ignored or asked for reassessment - in real-time. With the acceptable assessments, a group assessment can be generated and used to determine the optimal CAs.

The choices, that is, the recommended changes in planned operations, by each of the experts whose assessment was found to be acceptable, as well as the new CAs based upon the aggregated group assessment can be presented in graphical form to the experts. An aggregation of the accepted CAs could also be done. For example, if a majority of the experts agree on one CA, this CA could be seen as the aggregated CA which could be compared to the CAs resulting from the aggregated assessment. As will be discuss later on, the two do not have to be identical, even if all experts agree on one CA.

The operational risk management view of the multi-expert assessment and choice process is as follows.

- Each expert makes an assessment of risk, cost, and benefits for the activities affected by the RTE (even for those activities which are not part of any operation, i.e., CA). It is assumed that the affected activities have been identified by one expert (e.g., operational risk manager) and submitted to all other experts. If there is no agreement on the affected activities, there can be no consideration of aggregation of assessments. However, the choices, that is, the CAs recommended by the experts, will still be valuable.

- Whenever an expert makes an assessment, the assessment is evaluated for the purpose of determining a group assessment. A weight is determined for each expert; this weight is a function of (i) his/her consistency in past assessments, and (ii) his/her degree of group membership. The possibility that any expert gets excluded or that one expert could become the sole expert (i.e., his/her weight becomes one) should be avoided. This is called the *recognition axiom* in social choice literature [Arrow, 1951]. That is, all experts are recognized.

- The weights of the experts and their assessments are used to determine the acceptability of the assessments. This may result in some assessments being rejected, or some experts being asked to do a reassessment. The reassessment process can continue, depending on the time available for decision making. It should be noted that a rejected assessment does not mean that the expert was not considered. S/he still contributes to the consistency of the group and thus influences the group assessments for subsequent RTEs.

- A group assessment is computed using the weights and the acceptable assessments.

- For each expert and for the group assessment, the changes in CAs are computed and the results presented for decision making.

Preference aggregation for multiple experts has been addressed both in the context of risk management and in the context of social choice. In the latter case the issue has been the subject of research for more than 200 years and resulted in well known paradoxes, such as rank reversals in the *Borda count,* the *Condorcet paradox* of non-transitivity, and *Arrow's impossibility theorem*. Several methods have also been proposed for the combination of forecasts [Brodley, 1982]. As far as the aggregation of experts probability distributions is concerned, several algorithms have been proposed [Clemen, 1989]. For practical purposes, simple averages of point estimates have often shown to suffice, although they have some conceptual drawbacks. Recent developments in the aggregation of experts' opinions in risk management include research by Beinat et al. [1994], Cooke [1991], DeWispelare et al. [1995], Sandri et al. [1995], and Myung et al. [1996].

The proposed multi-expert preference aggregation approach is a dynamic real-time process that exhibits adaptive behavior which holds for both conditional and non-conditional preferences. The relationship of an individual expert's assessment to that of the group, as well as how an expert's assessment varies over repeated assessments, will change, and may adapt to group norms. For example, if one expert is consistently the most conservative, s/he should not be "punished" for being so. If experts are presented the results of their assessments relative to the other experts' assessments, they might undergo a (e.g., learning) process and change their attitude toward risk and costs and, subsequently, toward assessment and choice. The models presented in the following sections allow for adaptive behavior by the experts.

3.2 An Adaptive Model for Cardinal Preference Assessments

Cardinal assessments are done for operational costs and risks for the multiattribute utility (MAU) model discussed in Section 2.4 of this chapter. With n experts, e_i, $i=1,...,n$, where for the k-th RTE (stage k), RTE_k, each expert performs a cardinal assessment, $v_{i|k}$. The aggregated group assessment for the n experts is defined as the weighted average, where $w_{i|k}$ is the **weight** of expert e_i in stage k:

$$m_k = \left(\sum_{i=1}^{n} w_{i|k} \right)^{-1} \times \sum_{i=1}^{n} v_{i|k} w_{i|k}.$$

It should be noted that the weighted mean value is invariant over multiplicative transformations of the weights. In other words, if the weights are rescaled by multiplying them by a constant, the weighted mean is still the same.

The weight of expert i in stage k, $w_{ilk}=f(c_{ilk-1}, d_{ilk-1})$, is determined iteratively and is a function of the expert's long-run consistency, c_{ilk}, and his/her **relative deviation** from the aggregated group assessment in the previous assessment(s):

$$d_{ilk} = \left| \frac{v_{ilk-1} - m_{k-1}}{v_{max|k-1} - m_{k-1}} \right|.$$

The expert whose assessment was furthest from the aggregated group assessment in stage $k-1$ has in stage k a deviation coefficient, $d_{ilk}=1$ (least agreement with the group). If there is an expert whose assessment was equal to m_{k-1} then the deviation coefficient is $d_{ilk}=0$ (highest agreement with the group). The **long-run consistency**, c_{ilk}, is defined as the long-term average rank change:

$$c_{ilk} = \frac{1}{(n-1)(k-j)} \sum_{j=1}^{k} r_{ijlk},$$

where $1 \leq j \leq k-1$, with j being the first stage to count from. The maximum value (least consistent) with n experts is $c_{max|k}=1$ and the minimum value (most consistent) is $c_{min|k}=0$, if the expert never changes rank. The rank change, r_{ijlk}, is defined as the number of positions an expert changed compared to the previous assessment. For example, let's assume that an expert provided in stage $k-1$ the second most conservative assessment and in stage k the fourth most conservative assessment; in this case, the rank change is $r_{ijlk}=2$. The stage j is determined by the total number of past assessments of RTEs by the group. For example if daily traffic congestion results in assessment and the rerouting of trucks carrying hazardous materials by the same group of experts, a large j (e.g., $j=k-1$) may be selected to represent the most recent assessments by the group.

The weight for each expert can now be computed as a function of the expert's long-run consistency, $c_{ilk} \in$ [0 (most consistent), 1 (least consistent)], and his/her relative deviation, $d_{ilk} \in$ [0 (in highest agreement), 1 (in lowest agreement)]: $w_{ilk}=f(c_{ilk-1}, d_{ilk-1})$. To compute the resulting weight, an approach is proposed which was first introduced as the Minkowski metric and then extended to fuzzy logic by Yager [1980] and employed in the Swiss safety regulation for disaster scaling [BUWAL, 1991]. The **weight** for expert i in stage k is:

$w_{ilk} = \text{Round}(10 \times w^*_{ilk} + 1)$, where $w^*_{ilk} = 1 - \min\left[1, \left((c_{ilk-1})^\kappa + (d_{ilk-1})^\kappa\right)^{1/\kappa}\right]$,

and $\kappa = 1, 2, \ldots$, and $w^*_{ilk} \in [0,1]$. The characteristic of the parameter κ is that for $\kappa = \infty$, $w^*_{ilk} = 1 - \min[1, \max(c_{ilk-1}, d_{ilk-1})]$, and for $\kappa = 1$, $w^*_{ilk} = 1 - \min[1, c_{ilk-1} + d_{ilk-1}]$; i.e., $w_{ilk} = 1$ only if $c_{ilk} = 0$ and $d_{ilk} = 0$. The expert's weights are therefore $w_{ilk} \in [1, 11]$.

To determine the acceptable assessments, a **range of acceptability** (δ_k) must be determined. This range of acceptability is computed for the n assessments provided in stage k using the weighted mean value. We propose to do this as a function of the unbiased sample variance (s_k^2) of the weighted assessments:

$$s_k^2 = \frac{\sum_{i=1}^{n} w_{ilk}(v_{ilk} - m_k)^2}{\sum_{i=1}^{n} w_{ilk} - 1},$$

which converges toward the variance under multiplicative transformation, that is:

$$\delta_k = \sqrt{\lim_{w \to \infty} s_k^2}, \text{ where } \lim_{w \to \infty} s_k^2 = \frac{w \sum_{i=1}^{n} w_{ilk}(v_{ilk} - m_k)^2}{w \sum_{i=1}^{n} w_{ilk} - 1} = \frac{\sum_{i=1}^{n} w_{ilk}(v_{ilk} - m_k)^2}{\sum_{i=1}^{n} w_{ilk}}.$$

The range of acceptability is a function of the **group consistency**, C_k, which gets computed at each assessment stage. The dynamic group consistency is a measure of disarray of the rank-orders of the experts' assessments over time. This measure assumes that differences in the assessments are less serious if they occur because the experts have different degrees of conservatism. Assume that there are three experts and that one consistently assigns highest risk values, another always lowest, and the third lies always in-between. Then, the group as a whole is consistent and one would be more willing to accept differences between the experts than in the case where the preference-orders change very often.

The group consistency measure is therefore a relative measure of concordance of conservatism or risk attitude. Different measures have been proposed in the literature with the most prevalent one being Kendall's coefficient of concordance, which is used to test rank correlations [Kendall and Gibbons, p. 119, 1990]. Thus, the group consistency measure is defined as:

$$C_k = \frac{S}{S_{max}} = \frac{12S}{n^2(k^3-k)-n\Sigma U'}, \text{ with } S = \sum_{i=1}^{n} R_i^2 - \frac{kn^2(k+1)^2}{4},$$

where the R_i^2 are the squares of deviations of the rank sums around their mean which equals $k(n+1)/2$, and $U' = (n^3 - n)/12$ is to reduce the sum of square of deviations due to tied ranks. For example, assuming that two experts assess the cost as 66, three experts as 72, and another 3 experts as 69. Then, the correction factor is $U'=[(2^3-2)+2(3^3-3)]/12 = 4.5$.

The group consistency is a measure of relative agreement of k rank-orders. It takes on the value $C_k=1$ for perfect agreement, and the value $C_k=0$ for complete disagreement. Consequently, the feasibility range is defined as:

$$v_{ilk} \in [m_k - \gamma_k C_k \delta_k, m_k + \gamma_k C_k \delta_k],$$

where γ is a constant. For $\gamma_k=1$, the feasibility range is the standard deviation multiplied by C_k. A small value of γ_k causes many assessments to be rejected. This would be reasonable if time is available to have the experts who fall outside the acceptability range do a reassessment. Moreover, γ_k could vary in with increasing k.

Instead of using a behavioral approach where some of the experts might be asked to do a reassessment, the acceptable assessments at each stage can also be determined mechanically in an iterative way. First, with all n assessments, the range of acceptability is computed for some γ. Assessments falling outside of this range are disregarded. With the remaining assessments, a new range is determined by computing new m_k and δ_k values with these feasible assessments. Then, the unacceptable assessments are discarded. This procedure is repeated until all remaining assessments are contained within the range of acceptability (i.e., until none of the remaining assessments is rejected). However, if at the first iteration the acceptable set is empty, the assessment is not valid and the decision maker must rely on his/her own assessment. Thus, the algorithmic approach of determining the acceptable assessments is as follows:

```
F = {v_ilk}, (set of all assessments), j=0
repeat while F_j' ≠ Ø (the set of rejected assessments)
    j=j+1
    compute m_klj, C_klj, and δ_klj,
    {v_ilk} ∈ F → F_j', where v_ilk ∉ [m_klj - γ_k C_klj δ_klj, m_klj + γ_k C_klj δ_klj]
    F = F \ F_j'
    if F = Ø and j = 1 then assessment is bi-modal and unacceptable
end repeat
```

Note that the set of acceptable assessments could be empty for two cases: (1) if the group assessment for an RTE is bi-modal, and (2) if the experts are so inconsistent over time such that C_k and δ_k are very small, resulting in a range of acceptability too small to include any assessment.

3.3 An Adaptive Model for Ordinal Preference Assessments

The assessment spectrum for an ordinal risk scale, Π, was introduced as consisting of several preference classes referring to cost and risk attributes and also to the priorities between the two (the avoidance of HR actions has higher priority than of C actions, and higher priority than of LR actions): $\Pi = \Pi^{(1)} < ... < \Pi^{(n)}$. A possible assessment spectrum for the OP model is:

$$\alpha < HR < C < LR < \omega.$$

Assume that the attribute-classes for the risk attribute (in this example the two classes HR and LR), as well as the classes α (action must be avoided) and ω (impact of RTE is negligible) consist of only one element, and the classes of costs of multiple elements. It is further assumed that an assessment is done properly if every activity is assigned one element reflecting every attribute. For ORM, an assessment is **proper** if every activity which gets assessed has a risk-element from $\{\alpha, HR, LR, \omega\}$ and a cost-element from $\{\alpha, \{\$\}, \omega\}$.

With n experts, the probability that all experts assign the same risk value is $4^{(1-n)}$. If their assessments were independent, the probability of having r identical assessments out of s assessments, where $r \leq s$, would be binomial distributed, where $p = 4^{(1-n)}$:

$$p_s^{(r)} = \binom{r}{s} p^s (1-p)^{r-s}.$$

For example, the probability that eight experts agree at least once in 18 assessments is about only 10^{-3}. Thus, it is never expected that all n experts agree on all r assessments. On the other hand, with four risk classes and n experts, $trunc(n/4)$ experts must agree on at least one out of the four risk classes. A group assessment is therefore called **sufficient** if a majority of the (weighted) experts agrees on the tradeoff between risks and costs. For the assessment spectrum $\{\alpha, HR, LR, \omega\}$, this means that the assessment is sufficient if the majority of experts agree on one of the two sets {α,HR} or {LR,ω}. The corresponding set of assessments is then referred to as the sufficient set.

An assessment is called **efficient** if it is sufficient and if a majority of the experts choosing the sufficient set agree on one preference. The corresponding set of assessments is referred to as the efficient set. Finally, an assessment is called **satisfactory** if a majority of experts agrees on one preference class. The corresponding set of assessments is referred to as the satisfactory set. Evidently, a satisfactory assessment is efficient. To see this, let $\Pi_<=\{\alpha,HR\}$ and $\Pi_>=\{LR,\omega\}$. Let Π_i be the preference class that determined that the assessment is satisfactory; that is, by definition $|\Pi_i| > |\Pi_j|$, where $i \ne j$. Since $\Pi_i \in \Pi_<$ or $\Pi_i \in \Pi_>$, it follows that the assessment is efficient.

Acceptability is defined to include either efficient or satisfactory group assessments. Therefore, there may be a sufficient assessment which is not acceptable; for example, two experts assess α, two experts assess HR, one expert assesses LR, and two experts assess ω (where all experts have equal weight).

An RTE may cause several actions to be reassessed. So far, only the concept of how much multiple experts "agree" on the preference assessment of one action has been introduced. However, a measure is needed to account for consistency in rank-order. An expert that consistently assigns lower preferences to links is conservative; that is, the higher the risk the lower the preference for that link. Being conservative is acceptable, but if the group of experts has different experts assigning low preferences, the assessment is not consistent.

Therefore, a group assessment must be defined equivalent to the mean group assessment for the MAU model in Section 2.4. If a group assessment is satisfactory, the aggregated assessment is the preference class that is agreed upon by a majority of the experts (e.g., at least five out of eight experts). If a group assessment is efficient, the aggregated preference is the preference class that receives a majority within the two classes that are acceptable. For example, if (for equally weighted experts) two experts agree on α, three assess HR, one expert assesses LR, and two experts assess ω, we have an efficient, that is, acceptable, assessment, with HR being the aggregated group assessment. It should be noted that we do not use the class with the highest number of assessments as the group aggregated assessment. For example, in a case involving nine experts, four experts might assess LR, zero ω, three HR and two α. Then, the group aggregated assessment would not be *LR*, although it has the most experts (but not a majority) agreeing on this class, but HR because the assessment is efficient.

The distance measure to the aggregated assessment can now be defined. The non-efficient values of an efficient assessment have the distance measure $d_{ilk}=0$. The efficient values which do not fall into the class which determined the assessment to be efficient have the distance measure, $d_{ilk}=0.5$. Finally, the assessments that fall into the group aggregated class have a distance measure, $d_{ilk}=1.0$. It should be noted that a non-efficient assessment does not provide an aggregated group assessment. In such cases, the decision maker must rely on his/her own assessment.

Using these distance measures and the consistency coefficients of the experts, the weights are updated for the assessment of the next RTE. The aggregation of the assessments does not, however, need to be done by an iterative procedure because the process of assessing the efficient set already takes care of the concept of eliminating outliers. One might argue that the group aggregation procedure proposed for ordinal assessments is more severe than the one for the cardinal assessments because of the different definitions of acceptability of the group as a whole. A cardinal assessment of a group is not accepted if the first iteration eliminates all assessments. An ordinal assessment is not accepted if it is not efficient, although it might be sufficient. The reason to allow this more severe definition is based on empirical evidence which shows that the ordinal assessment outperforms the cardinal assessment in ORM situations in terms of effort and accuracy (see Chapter VII). Thus, it is expected that the ordinal assessments are more consistent than the cardinal assessments, especially when one attribute is assessed on a cardinal scale and the other is assessed on an ordinal scale.

3.4 Choice Process

The decision maker has the choice of maintaining the present course of action or creating a new set of CAs in response to RTEs. In the case of ORM the decision maker may be required to or may need to call upon "experts" to provide recommendations. This process was modeled as first one of assessment and then as one of choice. However, in the case of changing CAs the choice phase is performed algorithmically based upon the assessment. Therefore, the decision maker has both the planned CA and a set of recommended CAs, which may include the planned CA; that is, the CA based upon the group assessment and the CAs based upon the individual expert assessments' for those experts whose assessment was found to be acceptable. These recommended CAs can be displayed graphically. In addition, the decision maker can create his/her own CA.

Two CAs, R_i and R_j, are referred to as **indifferent** (preferentially equivalent) if they have the same overall preference: $\pi_{R_i} = \pi_{R_j}$. They are called **congruent** (strategically equivalent) if they consist of the same links in the same sequence: $l_i^{R_1} = l_i^{R_2}$. They are referred to as **analogous** if they are congruent but not indifferent. And, finally, they are called **identical** if they are congruent and indifferent.

With any given assessment of the affected activities (e.g., the experts' assessments or the group aggregated assessment), alterations to the planned CAs can be computed. The **feasibility** of CAs can be defined in terms of how many experts come up with the same CA. In general, there are only two CAs possible with decision making by attribute – the planned CA and the CA which avoids the RTE. Thus, some experts will propose the alternative CA while the others will not. The aggregated assessment results also in one of the two CAs. The number of experts choosing the planned CA and the CA resulting from the aggregated assessment can now be compared.

The aggregation of multiple expert assessments for decision making, however, could result in counterintuitive choices. Simple numerical examples can be constructed for which the linearly aggregated expected utilities (i.e., the risk) of two experts result in a choice that contradicts both experts' assessments. An assessment aggregation procedure should avoid this rather awkward result. Social choice researchers have thus proposed the *Pareto optimality* axiom which says that if all experts agree on one alternative, then the aggregated assessment should result in this alternative [Arrow, 1951]. Unfortunately, aggregation principles based on averages do not necessarily comply with this axiom.

To see this, let's assume there is an RTE which affects one action of the planned CA_1. A change of the planned CA_1 to CA_2 is suggested if the newly assessed risk and/or cost values for undertaking this activity are too high. All risk-cost pairs which result in keeping the planned CA_1 define a convex set. Figure 3 shows on the left the convex set of velocity-risk pairs that will result in remaining on the planned CA_1 (where velocity reflects the travel costs). If E_1 and E_2 are two of such assessments provided by two experts, then the aggregated assessment (E_A) will lie on the connecting line between the two assessments, which, by definition, also means staying on the planned CA_1 (that is, not avoiding the actions affected by the RTE).

The right hand side of Figure 3 shows that there could be two assessments that lie outside the convex set (i.e., which suggest changing the planned CA_1 to CA_2). However, a linear aggregation of the assessments might result in an aggregated assessment leading to a CA which will contradict both experts' choices (i.e., not to avoid the RTE while both experts suggest avoiding it).

Because of the possible violation of the Pareto optimality axiom, all alternative CAs based on the assessment of the "acceptable" experts, as well as the alternative CA resulting from the aggregated assessment should be presented to the decision maker. The operator can then decide which of these alternative CAs to follow. If all the experts' assessment result in the same CA, one that is different from the CA resulting from the aggregated assessment, a reasonable heuristic would be to choose the former CA. In any case, the *recognition* axiom should always be considered, saying that all experts must be considered in the aggregation of the assessments. In terms of the proposed models, this simply means that no expert will ever receive a weight of zero, a situation which would exclude him/her from the group.

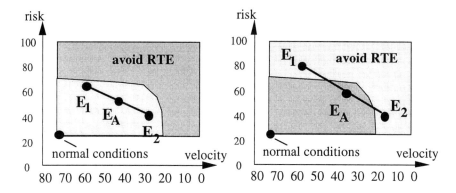

Figure II.8: Violation of the Pareto optimality axiom.

The following chapter reviews advances in information technology. Embedding the formal constructs discussed in this chapter into these technological advances provides the basis for decision support for operational risk management. Examples of decision support systems for different fields of operational risk management are presented in Chapters IV, V, and VI.

CHAPTER III: ADVANCES IN INFORMATION TECHNOLOGIES

Advanced communications technologies (as discussed in Chapter I), together with the latest advances in information processing technologies (as will be discussed in this chapter), provide the technological basis for decision support of operational risk management. This chapter discusses three advances in information technology: multimedia and hypermedia, the Internet system, and virtual reality. To build an intelligent decision support system for operational risk management, information and communications technologies must be complemented by decision models for operational risk management, as was discussed in Chapter II.

1. Multimedia and Hypermedia

1.1 Multimedia

While research has focused on individual modes of media presentation and decision support systems (DSSs), such as video conferencing [Chidambaram and Jones, 1993], little attention was given to the integration of multimedia and DSSs [Burger, 1995], [Hatcher, 1995], [Ramesh and Sengupta, 1995]. Integrated multimedia presentations can focus on the information the presenters want to convey much more effectively than a single type of media can do. Consequently, a distributed multimedia system for dispersed groups of operators, managers, and experts provides important information about a situation. It can evaluate the information and collect and distribute judgment and knowledge from and to various individuals and groups. This provides all members of the decision making organization with the information and knowledge needed for common understanding of the situation. Therefore, their recommendations can be based on this understanding, avoiding unnecessary information processing.

A) Multimedia on the Network

Multimedia requires a large amount of space for data, making compression and decompression (coded) technology extremely important to send/receive multimedia data on a computer network. For example, a 640x480 picture with 24-bit color (millions of colors) needs 7.4M bits (920 kB); one minute of the same size and color digital movie needs 13.32 G bits (1.66 GB); and one minute of Compact Disk (CD) quality sound needs 10 MB. Ethernet local area network (LAN), which is the most popular LAN, can transfer data at a speed of 10 Mbits per second (bps). A one minute digital movie needs 22 minute to transfer on an Ethernet LAN.

To construct a real-time system on a computer network, coded technology must be employed. Time-dependent dynamic presentation, as opposed to still media, has the most impact on multimedia networking. Time-dependent media includes digital video, voice, live music, and live video.

B) QuickTime

QuickTime (a trademark of Apple) is based on the concept of standardizing a digital movie on a personal computer (Macintosh, Windows), including common user-interfaces, video/audio coding, file format, and application programming interfaces (API). In addition to video (or time-based data), the standard also applies to still media, such as pictures, and it integrates time-based data into mainstream applications, including multimedia authoring tools. QuickTime manages video and audio in separate tracks, and each track has a timetable, which makes the synchronization of video and audio possible.

QuickTime Conference simplifies the exchange of time-based data among computers connected with a computer network, using QuickTime's coded method and its standardization, in addition to Apple's AppleTalk (a trademark of Apple) protocol and TCP/IP computer networking. The programming components of QuickTime Conference manage the user-interfaces, conference events, transportation of the conference data, initializing connections, etc. In addition to conference data, programmers can exchange their data, such as video and sound. With these functions, vital data for multimedia communications of multiple groups of experts can be sent and received. Because QuickTime Conference can be used on either LAN or TCP/IP, a system can be connected with either LAN or TCP/IP, and the connection is easy to switch back an on. QuickTime Conference technology was use to develop the decision support systems and to integrate the decision logic for the multi-group emergency management layout, as will be discussed in Chapter VI, Section 2.

1.2 The Hypermedia Environment

A) Basic Concept

A multimedia authoring tool was chosen as the environment for developing the prototype systems which will be discussed in Chapter V. The underlying concept of multimedia systems is hypermedia which allows connections to objects in a network structure. Hypermedia is a powerful concept that has changed the organization of information and procedures (objects). It allows the integration of text, graphics, audio, and video in decision support systems. It also has the advantage of being easily extensible for user-specific purposes. While objects are traditionally processed in a sequential way, the hypermedia concept is based on a graph structure. The underlying programming concept is referred to as Object Oriented Programming (OOP), and the editing environment is known as Hypertext. The nodes (objects) can contain text, graphics, executable programs, or any other form of multimedia information and procedures (e.g., video and sound).

The oriented links represent relationships between two nodes, which can consist of sending a message, linking nodes, activating a successor node, "jumping" to another node, or any other relation. OOP has advantages over sequential processing. It offers more flexibility and power in the development as well as in the execution phase. Hypertext and hypermedia have been the subject of research, writing, and experimentation for more than 20 years [Yankelovich et al. 1988].

Different multimedia authoring tools are commercially available [West, 1993]. The best known is probably HyperCard by Apple. Comparable multimedia environments are Supercard, WinPlus, MediaObject, and Director. They all incorporate a spoken-like object-oriented programming language. Five object classes are the standard for these types of systems: button, field, background, card and stack. Objects have properties and scripts. For example, properties of the object class window are text style, size, font, etc. Scripts are program codes (also called message handlers) that can receive messages from other handlers, the mouse, or the keyboard. They can also send messages to other message handlers. A message handler that is activated by the computer mouse starts with "on MouseX" and ends with "end MouseX." The X stands for any mouse activity, such as "up," "down," or "within." For example, the handler

```
on MouseDown
    computeRoute
end MouseDown
```

computes the optimal route, using the message handler computeRoute, whenever the mouse is clicked within the area of the object. A handler can also contain messages that can be sent to other handlers for execution. The addressed message x is the following: "send x to y of z," where y is the receiving handler and z the object holding the handler. Unaddressed messages (e.g., "computeRoute") go through a predefined object class hierarchy as shown in Figure III.1.

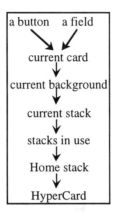

Figure III.1: Object class hierarchy.

This message structure implies that for multiple identical handlers at different hierarchical levels, only the handler at the highest level processes the unaddressed message. The message structure is an oriented graph with default arcs along the hierarchy. Thus, the object oriented structure of hypermedia environments is a super-graph built around the default graph structure.

Handlers are very versatile and can execute basically anything, including computations, animations, voice output, and video. In addition, predefined standard procedures for animation and computation are available from public domain systems.

Figures and scanned photographs are best put on the background level, while animation and drawings should be placed in the foreground. In addition, figures and pictures can be shown in different windows, and predefined visual effects can be chosen from a standard library. Animation is done either by rapidly changing the graph of an object or by moving the object to different locations along the coordinate system. Finally, the hypermedia system can be transferred to a stand-alone application which facilitates portability to other computers.

Hypermedia environments, as opposed to traditional programming languages, are well suited for rapid prototyping by analysts who are more interested in the conceptual *what* then the technological *how* of developing a decision support system (DSS). For further material on hyper programming see, for example, [Coulouris and Thimbleby, 1993].

B) The Programming Environment

The spoken-like programming language is very powerful and has the advantage that the characteristics of variables need not be defined explicitly. For example, the script for the sum (x) of two values $a_{5,3}$ and $b_{4,2}$ of two arrays (called fields) A and B is: {put item 3 of line 5 of card field A + item 2 of line 4 of card field B into x}. Although the handlers are compiled in memory at their first activation, execution speed is rather slow. To overcome this limitations, computationally intensive procedures can be written in Pascal or C and be included as external commands (XCMD) or external functions (XFCN), or the codes can be compiled with a special compiler.

The multimedia authoring tools provide a fully developed programming language called Hypertalk (for Hypercard) or Mediatalk (for Media object), which can also read and write data to other files or applications. The modular structure of the programming environment and the available high-level commands significantly cut down the development time and reduce the chances of error. If execution speed is critical, procedures with extensive number crunching are compiled in lower-level languages and integrated in HyperCard as external commands (XCMD). Some authoring tools allow direct compilation of their scripts. It is also possible to execute other applications from within HyperCard, such as routing algorithms written in C or Pascal.

Multimedia environments are especially well-suited for control tasks. Data and codes can be transmitted and received through the serial port [Bond 1988]. These capabilities provide the base for easily connecting decision support systems to sensors monitoring the environment for any possible real-time event. Other advantages of developing a decision support system on the personal computer level are that hardware costs are low and the DSS is highly portable. This makes testing and future implementation of the DSS in companies' headquarters easier to achieve.

Another advantage of hypermedia is that Hypertalk shells are now available, which allow the user to add expert system reasoning easily to any program developed in the hypermedia environment. Payne and McArthur [1990] see HyperCard as the future of scripting languages. They programmed the expert system Entrypaq entirely in Hypertalk, so that users can examine how this expert system shell works and implement it in their own programs. Another way to incorporate artificial intelligence reasoning into HyperCard is via DELPHIA PROLOG (also known as HyperProlog), which uses HyperCard as a graphical front-end.

Marchioni and Schneiderman [1988] see the importance of Hypertext systems in their potential capacity to augment and amplify human intellect and to overcome human limitations. What makes this possible is that the designer can more easily tap into the basic cognitive processes that guide information seeking. The so-called direct manipulation interfaces (e.g., mouse, touch-screen), which can easily be incorporated into HyperCard applications lead to less cognitive load in using computers. As a result, users can concentrate more on the task at hand and rely less on their limited working memory.

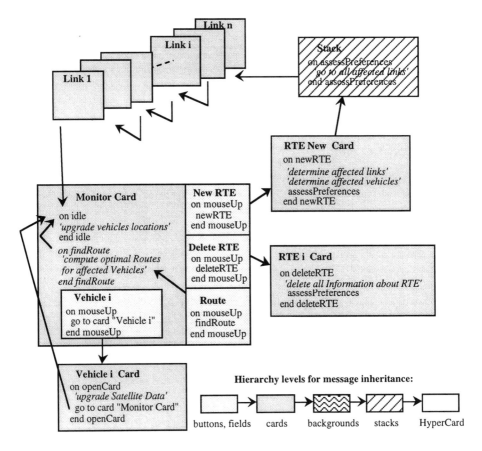

Figure III.2: The basic hypermedia structure of the DSS.

The prototype DSSs which will be discussed in the subsequent chapters have been developed within a so-called stack. A stack is a collection of several cards which can be added continuously to the stack. Several cards can share the same background. Each card can contain button and fields that are technically objects, able to handle and send messages. In addition, graphics can be added to each card, either on the background or on the foreground. Background objects and graphics apply to all cards that share this particular background, while card objects and graphics pertain only to a specific card.

Figure III.2 shows the basic graph structure for a possible DSS for operational risk management. The names in bold indicate objects at different levels. The objects on the top level are the buttons 'New RTE,' 'Delete RTE,' 'Route,' and 'Vehicle i.' The objects on the next lower level are the cards 'Monitor Card,' 'RTE New Card,' 'RTE i Card,' 'Vehicle i Card,' and the n link cards. The object on the lowest level is the 'Stack.' Each object has at least one script written in Hypertalk. In addition, some scripts use external commands. The scripts start with "on message" and end with "end message." The message can refer to system messages, for example, the mouse input, such as "mouseUp," "mouseDown", mouseStillDown," and "mouseWithin," or to user-defined messages, such as "assessPreferences," "newRTE," and "deleteRTE."

C) Systems Architecture and Functionality

The hypermedia system consists of four components: navigator, interface, modelbases, and databases (see Figure III.3). The core of the system is the navigator. In the case of the transportation of hazardous materials, its tasks are to assign routes to the vehicles, to generate events, to control the clock, and to manage the interface. Such a combination of routes and events affecting the shipments is called a scenario. The system can therefore be used by operators to simulate decision making for the four modeling paradigms and with different scenarios.

The modelbases consist of an algorithm to identify events (i.e., the links of the road network that are affected by an event, for example a snow storm), an algorithm to identify the vehicles that are affected by an event (i.e., the vehicles that plan to drive through the area that is affected by the event, for example, the vehicles that plan to drive through the snow storm), different models to assess the impact of the events on transportation safety and efficiency, and routing algorithms.

Different modeling paradigms imply different models. For example, the assessment and the routing models for a utility paradigm are different from the models for an ordinal preference paradigm. Each model and data file is stored, in terms of hyper programming, on a card. Then, different models for the same task (e.g., all the routing models) are grouped into a stack. Therefore, the assessment and routing models are grouped into the stack "routing of vehicles," while the models for the identification of events and affected vehicles are the same for all four modeling paradigms.

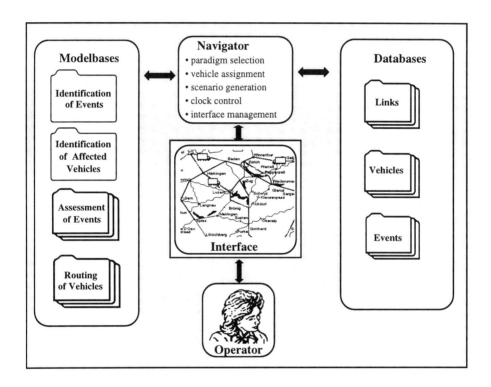

Figure III.3: Architecture of the hypermedia system.

The databases are grouped into three stacks. The link data base contains information about every road segment, such as the length of the link, the coordinates, and the risk and cost values of traveling along the link. Because the area of Switzerland which is currently incorporated into the system contains 76 links, there are as many cards in the link stack. The vehicle stack contains nine vehicles, and the event stack contains six events. A combination of vehicles and events makes up a scenario.

The user communicates with the system through the interface. The navigator controls the interface. This means that the navigator manages the animation (e.g., movement of vehicles on the map), voice output, display of scanned photographs, and user data input.

2. Internet

2.1 Introduction

Technological advances in communications, information, and computing technologies are revolutionizing the daily work of managers. In a few years, the Internet system will cover the whole world, with hundreds of thousands of service providers and millions of users [Borsook, 1994]. All types of digitized code can be transferred via high-speed digital networks in asynchronous transfer mode (ATM).

Satellite systems are commercially available for world-wide real-time communications and positioning of remote and mobile units. They allow data, voice, and video communications, as well as other services, such as remote access, video conferencing, and real-time monitoring of vehicle fleets.

Computing technology has matured to include multimedia computer systems with audio and video input and output. Hypermedia and virtual reality are being considered to support decision making in emergency management as will be discussed further in Section 3.

The combination of advanced communications, information, and computing technologies results in a desk-top multimedia computing system as illustrated in Figure III.4. In a multi-finder system, such as provided by X-Windows under UNIX, the Macintosh system, or Microsoft Windows, multiple applications can run simultaneously. Text e-mail runs next to real-time data acquisition (e.g., for a real-time vehicle tracking system), and video phone (e.g. for a video conference and monitoring) runs next to interactive video (e.g. for a virtual reality application).

Systems for emergency management using some of these novel technologies have already been developed. An example is CAMEO, a PC-based multimedia emergency management information software system (MEMIS) for chemical production sites and transportation of hazardous materials [CAMEO, 1993]. Due to its user-friendliness, it has been implemented all over the world. Another example is InterClair, also a PC-based decision-support system, developed by the United Nations Interagency Program [InterClair, 1992]. InterClair assists with environmental modeling for risk assessment and management at all management levels. InterClair has been developed using the latest concepts, such as virtual instruments, animation, hypertext, and knowledge-based systems.

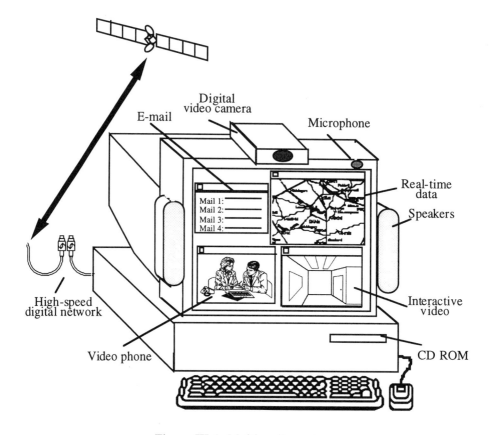

Figure III.4: Multimedia system.

2.2 The Architecture of an Extended MEMIS

The extended MEMIS discussed in this section was developed in a multimedia authoring environment. MEMIS provides an example of the technological support available for operational risk management. The hypermedia environment is based on the concept of object oriented programming. Objects at lower levels inherit the characteristics of objects at higher levels. An object has characteristics (attributes e.g., the size of a card or the shape of a button), and it can perform activities. These activities may be as simple as activating an audio or video message, or as complex as performing an algorithmic procedure that has been coded, for example, in Pascal.

MEMIS is built on a shell principle (see Figure III.5). At the heart of the system is the navigator, which controls the system's activities and communicates between the user (emergency manager) and the other levels. The first level, after the navigator, is the hypermedia environment. It has a prestructured architecture which supports the navigator. Codes written in the hypermedia scripting language can be compiled in memory. The code is attached to an object and can easily be altered. This is especially useful during the development phase. The designer of the system can also define different user-levels which would allow more in-depth access to the code. Thus, experienced users could have access to a lower user-level, while novice users would have access only to the higher levels. Moreover, code can either be compiled, or it can be written in Pascal or C and attached as external commands. Finally, stand-alone applications can also be generated which make the system independent from the development environment.

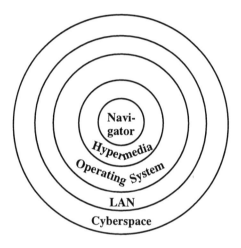

Figure III.5: Shell concept of MEMIS.

The third level is the operating system. Activities at this level are fired with an event handler, similar to the message handler in the hypermedia environment. This handler can start other applications, such as a word processor, and also control them to a certain extent. In other words, the navigator can communicate with other frequently used applications (as long as they are scriptable), such as word processors, spreadsheet programs, databases, drawing programs, etc.

The fourth level, the local area network (LAN), is basically identical to the previous level. The other work stations must, however, allow the navigator of MEMIS to access their file system and their applications. The major advantage of operating MEMIS on a LAN system is that applications and files can be shared among different users.

The last level is cyberspace, the world-wide information and communications network (Internet), based on phone lines, fiber optics, and satellite communications system. MEMIS attached to the Internet allows e-mail, video conferencing, real-time monitoring, remote database access, remote application access, etc.

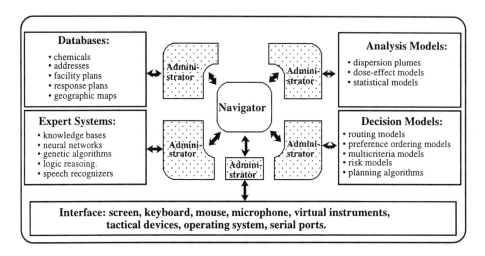

Figure III.6: Architecture of MEMIS.

At the functional level, MEMIS consists of databases, models, application programs, and expert system technology (Figure III.6). The core of the system is the navigator that facilitates communications between the human user, that is, the emergency manager, and the system's main modules. The communications interface between the navigator and the human user is based on different devices, including the computer screen, the keyboard, the computer mouse, microphones, virtual instruments such as slide bars and radio buttons, and, in the near future, tactical devices such as gloves. Communications between the navigator and the modules is supported by special administrators. An administrator activates the appropriate model or database, depending on what the task requires. The four main components of MEMIS are databases, analysis models, decision models, and expert systems components. Figure III.6 above shows the functional concept of MEMIS.

2.3 The MEMIS Modules

MEMIS consists of 25 different modules that are accessed or operate at different levels in the shell structure. While some modules are completely integrated into the hypermedia system, other modules reach all the way into cyberspace. Some of the modules are based on the ones used in the CAMEO system. The modules are shown in the navigator (see Figure III.7 below); a short description of the modules follows.

Memo

The Memo module is a note pad module where the emergency manager can place and retrieve notes. Coupled with the E-mail module, ticklers can be sent to other users.

Agenda

The Agenda module is a calendar which automatically notifies the user of appointments, meetings, etc. The notification time can be set minutes, hours, or days before the meeting takes place. Using the agenda module together with the E-mail module, agendas of remote users can be checked. Dates can be entered with notification and confirmation of the other user.

Addresses

The Addresses module contains the database of persons, institutions, emergency response teams, persons to be notified in cases of emergencies, etc. The Phone, E-mail, and WWW modules, along with phone, fax, and e-mail lines and personal home pages, can be accessed by double-clicking on the appropriate numbers. If the Addresses module is combined with the GIS (geographic information system) module, the location of the addresses on the geographic map can be shown.

Phone

The Phone module manages phone and fax messages. Together with the modem on a portable phone, MEMIS can be used as a portable system.

Help

The Help module assists the user in the use of MEMIS or in case of system errors. It can include video and audio instructions, as well as text.

E-mail

The E-mail module is used to communicate to other users and to receive messages from list servers.

Examples of list servers in emergency management are CMTS-L (chemical management and tracking; listserv@cornell.edu), DISPATCH (police, fire, and EMS telecommunications majordomo@comeng.com), EMERG-L (emergency services; listserv@vm.marist.edu), FIRENET (listserv@life.anu.edu.au), HELPNET (network emergency response planning; listserv@vm1.nodak.edu),

LEPC (hazardous materials emergency response; listproc@moose.uvm.edu), SAFETY (safety issues; listserv@uvmvm.uvm.edu). Moreover, governmental agencies run also list servers; such as the different EPA lists (e.g., EPA-Waste with all hazardous and solid waste documents; listserv@unixmail.rtpnc.epa.gov).

Figure III.7: Module of MEMIS.

Usenet

Usenet News is a service that manages articles prepared by people at educational, commercial, and government institutions all around the world. The articles are grouped into news-groups that focus on specific issues. The articles can be read with appropriate software by contacting a news server. Some of the NewsGroups relevant to emergency management are alt.disasters.planning, alt.med.ems, misc.emerg-services, sci.med.ems, and uiuc.safety (environmental health and safety forum). In the future, these articles will include graphics (displaying data), photo-graphs, and video, showing simulated situations and real disasters.

WWW

World-wide-web (WWW) is a hypertext client-server-based cross-referencing tool initiated by CERN. It includes file transfer protocol (FTP) and gopher. Further information about WWW can be accessed through anonymous telnet or ftp at info.cern.ch. An interesting WWW site for emergency management is, e.g.,

Global EMS Archives (http://herbst7.his.ucsf.edu). More information about related WWW, gopher, and ftp sites for emergency management can be obtained through ftp://hairball.ecst.edu/pub/ems/internet. emergency-resources.

Telnet

The module Telnet is used to connect to databases and on-line information services, e.g. libraries. In addition, telnet is used to access remote computer systems for, e.g., loading down accidents reports, weather reports, and historical data.

VideoConference

With a camera on top of the computer and the appropriate software (e.g., CU-SeeMe from Cornell University), video conferences and monitoring can be performed. Figure III.8 shows a screen view of a video conferencing session in MEMIS.

Figure III.8: Videoconferencing with MEMIS.

TextEdit

The module TextEdit connects to a common word processor.

TableEdit
The module TableEdit connects to a common spreadsheet program.

Calculator
The module Calculator connects to a calculator system.

GIS
The module GIS (geographic information system) connects to a geographic information system. This can be a commercial system or a task specific system such as that used in CAMEO. The internal system includes zooming capabilities, but does not have all the functionalities of a commercial GIS. This module can be connected to other modules, such as Phone, Addresses, Alarm, Emergency, etc. In addition, this module will include global positioning systems for the control of mobile units.

Alarm
The module Alarm dispatches warnings to emergency units and to groups affected by an emergency. It can be coupled with an automatic dialing telecommunications system. An example of an automatic dialing systems is the QuickCall telephone notification system used in the U.S. It is capable of simultaneously and automatically dialing around 1,000 numbers in 15 minutes.

Response Plan
The module Response Plan contains pre-planned response activities and evacuation procedures. It can be accessed during emergencies.

HazMat Databases
The module HazMat Databases contains data about the hazardous substances, such as physical state, level of concern, reportable quantity, etc.

Dispersion Model
The module Dispersion Model accesses dispersion models for the computation of diffusions. The models can be integrated at the hypermedia level or at the cyberspace level.

Scenarios
The module Scenarios can be used to devise response strategies. Dispersion plumes in stationary or dynamic systems can be computed.

Weather
The module Weather provides data on storms, local and regional weather conditions, etc. Its reports can be accessed by Gopher.

Accident Reports
The module Accident Reports is used to record and compile data on accidents for storage and reporting purposes.

Accident Statistics
The module Accident Statistics is a database with historical accident data, which can be purchased by professional accident statistics databases.

Installation Plans
The module Installation Plans contains the plans of installations. These includes floor plans, lay-outs of technological systems, and emergency escape routes. The plans are interconnected in a hypermedia system.

Inventory
The module Inventory is a database that supports the management of the facility. It tells what hazardous material is present and where it is stored.

Emergency
The module Emergency is used to manage the response to emergencies. It connects to other modules, such as Inventory, Response Plans, Addresses, Alarm, etc.

2.4 Tasks to be Addressed in MEMIS

The MEMIS system can be used for three tasks: daily work tasks, emergency planning, and emergency response.

Daily Tasks
MEMIS is a desk-top system that can be built around a workstation that is used in day-to-day operations in emergency management. It integrates the administrative tasks of an emergency manager with the emergency-specific tasks. Text processing, spreadsheet work, database management, and e-mail communications are performed in the same system as emergency management and planning. The advantage is that the user does not need to switch between two different system but rather can do all the work in one integrated system – MEMIS. With the access to cyberspace, the emergency manager can be kept up

to date about new developments. The workstation can be replicated in a laptop configuration for portability.

Emergency Planning

Scenario analysis, development of response plans, analysis of statistics, and so on, are routine tasks. In addition to these, the system can also be used for training and simulating actual incidents. The development of emergency response plans in a strategic setting can be done in a group decision setting. This can be accomplished by multiple experts gathering in a group decision support room, as shown in Figure III.9.

Figure III.9: Group Decision Making with MEMIS.

Emergency Response

This module includes capabilities such as automatic dialing, alarm, and real-time monitoring. Global positioning system (GPS) capabilities provide for real-time control of response resources. Routing models and expert systems provide recommendations to the emergency responders. CAMEO and similar models provide predictions of various impacts of the event.

2.5 Conclusions

MEMIS was developed in a prototype version to demonstrate how commercially available software and hardware could be integrated to provide emergency managers with a management information system, specifically designed for their needs. It shows that, at a relatively low cost – since most of the capabilities are already installed on personal computers – an emergency manager can have his or her own system, and not be dependent upon local, regional, or federal workstations and information systems. In addition, the system can either be installed on or duplicated in a laptop computer for portability and personal use.

The development of a MEMIS can be done by the emergency manager, in concert (if need be) with a person knowledgeable about PCs, word processing, and spreadsheets. The addition of modeling software, such as CAMEO, would require knowledge of dispersion models and traffic routing - topics familiar to emergency managers. It is certainly conceivable that every emergency manager could have his or her own MEMIS.

3. Virtual Reality

3.1 Introduction

The domain of virtual reality (VR) is not clearly defined. Moreover, similar terms are used for different kinds of concepts. Examples are artificial reality, virtual environments, teleoperation, virtual instrumentations, virtual interfaces, virtual worlds, virtual experiences, architectural walkthroughs, etc. What is common to all these concepts is that they aim at providing humans with synthesized sensations. The more realistic the perception, the more "virtual" the approach. Only genuine stimulation of human senses is considered VR; the creation of hallucinatory perceptions by ingestion, injection, or inhalation does not belong to the domain of VR.

If one attempts to bound the domain of VR, analogous philosophical questions arise as with the definition of artificial intelligence (AI) about 30 years ago. The question then was "When is AI achieved?" that is, when is a machine intelligent? A. Turing's answer to this question was that AI is achieved when a human working with an intelligent machine believes that s/he is working with another human [Tanimoto, 1987]. An equivalent question for VR can be asked: "When is a synthesized perception real?" To answer this, "reality" has to be defined. For some senses a definition equivalent to the one proposed for AI is introduced: virtual environments are real if one cannot tell that they are computer generated. To use this stringent definition for visual perception, however, would disqualify everything that up to now has been called VR, since technical limitations have thus far precluded visual simulations that are indistinguishable from reality.

It is important to distinguish between the goals of AI and the goals of VR. AI attempts to replace human intelligence by hardware, software, and data. VR, on the other hand, supports intelligence amplification (IA) rather than intelligence replacement [Rheingold, 1991]. The goal in VR and IA is to design systems that amplify, support, and stimulate the human mind by providing computer support in areas where machines outperform humans, such as computation, data storage, statistical inference, and deductive reasoning. In other areas, however, humans are superior to machines, and IA allows humans to continue to perform tasks which require judgment, pattern recognition, and intuition.

Pioneer work in VR originates from the entertainment industry. However, Morton Heiling's Experience Theater and Sensorama never got the credit they deserved [Rheingold, 1993]. Consequently, the entertainment industry did not promote VR concepts beyond Cinerama movie theaters. VR as it is known today, emerged from the field of computer science. Generally speaking, VR can be seen as the ultimate human-machine computer interface to an advanced simulation system. In fact, today's notion of VR consists of a human-computer interaction at various levels of reality, primarily focusing on visual perception. It is only recently that audio and tactile perceptions have been addressed in VR.

VR is an immersive, synthetic, or computer-generated environment that provides the experience of "being there" [Rheingold, 1991]. Ideally, VR is intuitive as well, allowing the user to communicate with the system via familiar and obvious actions [Wells, 1992]. In discussing VR, it is important to distinguish between natural and synthetic experience. In a natural experience, the user directly perceives the properties of something that is physically present before the perceiver. By contrast, in a synthetic experience, the user perceives a representation of something physically real rather than the thing itself [Robinett, 1992]. Examples of synthetic experience include flight simulation, robot teleoperation, everyday telecommunications, and, of course, virtual reality. Synthetic experience may be generated by the physical world, as in teleoperation (using virtual reality to perceive and manipulate objects remote in space from the user) or it may be generated by computer, as in architectural walkthroughs (using a VR version of an architectural blueprint, the user can "walk through" a building that does not exist in the physical world). Three levels of virtual reality have been identified: virtual space, virtual image, and virtual environment. Virtual space uses pictorial cues to represent a three-dimensional (3D) portrayal on a flat (2D) display. These displays commonly use perspective, shading, and textual gradient to create a "2 1/2D" image. Virtual images are perceived by adding various stereoscopic cues to produce 3D displays. Virtual environments add other sources of information with audio and tactile images.

3.2 Components of Virtual Reality

The basic component of VR is the human-machine interaction in a virtual space, also called cyberspace. In addition to the above mentioned output capabilities of VR systems, input devices through which the human conveys information to the machine add significantly to the sensation of VR. The ultimate goal for input devices is to reach the same realism as aspired for output devices; that is, voice, motion, and facial expression could be understood as readily by the machine as they are by another human.

A) Taxonomy

Robinett [1992] has proposed a taxonomy for classifying types of synthetic experience. Causality, model source, time, and space, are four dimensions on which synthetic experience can be classified. Causality refers to different ways that one can experience the world and breaks down into three categories: simulated, recorded, and transmitted models. In a simulated experience, such as a flight simulator, actions have effects in the synthetic world, but not in the real world. This distinction has important implications for emergency management training, where it is desirable for the trainee to experience the effect of his/her actions in the simulated environment, but where the consequence of those actions could be dangerous or unduly destructive if performed in the real world during a training exercise.

The second type of causality of experience is a recorded experience. A user who experiences a recorded event experiences what really happened, and the user's actions cannot change what happened. Recorded experience of a real-life emergency can be used to aid planning for future emergency response. An airplane's "black box" is an example of recorded experience. Transmitted experience incorporates the dimension of time, since the user experiences the events as they are happening, as in teleoperation, the robotic control of distant objects [Psotka and Davison, 1993].

The model source dimension refers to how the virtual world has been defined. Models may be scanned, constructed, or computed. For scanned models, data is extracted from the real world; night vision goggles produce a scanned model. With constructed models, elements are assembled piece by piece, as in cartoon animation, where each frame has been drawn by an artist, and then the frames are viewed in sequence. In computed models, simulation code generates the model data, as in algorithm visualization. For example, in a visualization intended to illustrate a bubble sort, the system simulates random data to be sorted.

The dimensions of time and space refer to the relative time and location scales of the user and the model. Models may be aligned, displaced, scaled, or distorted in time and/or in space. Aligned means that there is a one-to-one correspondence between the time/space of the user and the time/space of the model and that there is no change of location on the axis of time or of space. Examples include live TV, which is aligned in time, and night-vision goggles, which register the world in its actual dimensions. Models displaced in time or space have a one-to-one correspondence in scale, but at a different time or place; for example, a TV rebroadcast of a live event or teleoperation.

Examples of scaled models include slow-motion instant replay on TV, which is scaled in time, and microsurgery, which is scaled in space. For microsurgery, a one-quarter inch movement of the surgeon's hand may be translated into a one-hundredth of an inch movement of the instrument. Distortion of time or space differs from scaling in that distortion emphasizes or de-emphasizes one element of the model out of proportion to the rest of the model; for example, micro-teleoperation with exaggerated height or an edited TV broadcast.

B) Virtual Worlds and Cyberspace

A virtual world is a place that generates events that never really happen. It consists of a computer generated environment that the user interacts with. In more advanced systems, the user can even see him/herself acting in this environment. The use of computers coupled with automatic machinery to control and carry out complex operations is called cybernation. Consequently, virtual worlds are also called cyberspaces. One can not only move and act in cyberspace, but also touch objects of different textures, feel heat and pressure, look around in 3D, and hear sounds coming from different directions.

Cybernauts acting in cyberspace have a smaller radius of action than in the real world. While turning, seeing, and listening correspond to the actions one would do in the real world, moving can be done simply by pointing in the appropriate direction. Objects in cyberspace may be physical objects, or they may be artificial actors, that is, computer-generated representations of human forms, or even other cybernauts. One could shake hands with another person at a remote place and feel the pressure of the handshake. Meetings and conferences could take place in cyberspace. Emergency activities with different scenarios could be exercised in cyberspace. The response units would feel the heat, hear the noise, and see their actions being implemented.

Although these VR scenarios sound quite visionary, the technology to make them happen is becoming accessible, and time has come to investigate the potential use of VR scenarios for emergency management. Many concepts of VR have already been implemented in emergency management.

C) Technology in Virtual Reality

A major breakthrough in visual interactive input devices was reached in 1984 with the computer mouse and desktop system developed by Macintosh. The desktop represents a work place with disks, folders, and files as virtual objects. The mouse represents the virtual hand or finger which is used to point to those objects on the desktop. A single mouse click on an object selects the object and double-clicking opens the object. Objects can be moved around on the desktop and inserted into other objects. Since 1984, the mouse-pointing interface has been further developed to include virtual instruments, such as slide bars, knobs, toggles, and radio buttons. The input has become increasingly intuitive because the user knows from experience in the real world how to handle those virtual devices. Instead of using his/her hands and physical instruments, s/he uses the mouse as the virtual hand to work with the computer generated virtual input device, as illustrated in Figure III.10.

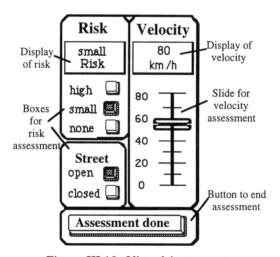

Figure III.10: Virtual instrument.

The latest desktop computer systems are equipped with remarkably effective voice recognition and voice synthesis features. Although there are not yet many software packages that can handle voice input, it will be only a matter of time until software houses start taking advantage of this novel input device. Soon managers will be talking to their word processor and spreadsheet programs.

The devices most frequently associated with VR these days are head mounted 3D displays and gloves. Three dimensional vision can be achieved only by providing the two eyes with different pictures that can be merged by the human brain into one 3D picture. This artificial 3D vision is called stereoscopic viewing.

More important than 3D perception from a motionless source, as is the case with 3D cinemas, is 3D vision that follows the user's head movements. This allows one to look around (or even walk around) the corner of an object in cyberspace. This 3D vision is achieved by head mounted cameras that keep track of the head position in space. In a somewhat more restricted way, this effect can also be achieved by 3D glasses coupled with an appropriate simulation system.

Gloves are used as advanced input devices. Sensors attached to the glove determine the relative position of the fingers and the glove in space. One can use the glove to move around in cyberspace, but also to pick up and move around objects. Advanced gloves include the possibility of tactile perception; that is, the glove can simulate pressure, heat, and the feeling of touching a rough or smooth surface. The ultimate goal in tactile output devices are suits that can act as artificial skin.

More advanced input devices are currently being researched. They include lip reading devices, gestural input devices, eye-movement detectors, and facial motion detectors. Anything the user does in front of the computer will be detected by sensors, used as input, and processed in the program. While so far only active inputs such as a mouse clicks can be perceived by the system, advanced computer systems will be able to recognize and process passive input; e.g., the facial expression for emotions. These advanced input capabilities coupled with the sophisticated 3D output systems will ultimately create what is known as mind amplifying computer technology - virtual reality.

D) The Role of Information Technology

The key to VR is advanced computing and communications technology. For every move or other kind of input, the simulation generates new graphics. Advanced microcomputer-based VR systems can generate from a couple of thousand texture- mapped polygons up to several thousand flat-shaded polygons per second. Power is provided today by the most advanced microcomputer systems, including RISC-based (reduced instruction set computers) microcomputers.

Data transfer between remote and mobile actors can be achieved by satellite-based phone systems, such as the Iridium system by Motorola. Eventually, at least 60 satellites will provide world-wide communications capability. In addition, fiber optics systems will transmit data at a speed equivalent to 50,000 typed pages per second. The Global Positioning System (GPS) and other satellite tracking and communications systems such as OmniTrac, EutelTrac, and Inmarsat C provide accurate positioning and real-time data transfer between any two locations on earth.

3.3 Impact on Emergency Management

VR has already had a significant impact on emergency management, including advanced data visualization systems within geographic information systems (GIS). Many advanced decision support systems (DSS) for emergency management rely on GIS technology and virtual instrumentation. Dispersion plumes are depicted on digitized backgrounds of aerial pictures and the growth of the plume is simulated and visualized in real-time. Concepts more closely related to today's notion of VR are virtual world navigation, tele-virtual conferencing, teleoperation, and simulation in cyberspace.

A) From CAMEO to Virtual World Navigation

A widely used emergency response DSS is CAMEO, developed by the U.S. National Safety Council. It originally was developed on a hypermedia system on a Macintosh computer. The advantage of hypermedia is that its object oriented graph structure allows easy navigation through the system. Hypermedia systems are often used as authoring tools for multimedia systems. Objects can be text, graphics, drawings, figures, or even environments. Hypermedia systems are used for a variety of applications, including simulation. Several examples of hypermedia decision support systems will be discussed in Chapters IV, V, and VI.

The concept of hypermedia has recently been used in conjunction with VR [Smith and Wilson 1993]. Instead of navigating through map displays as embedded into CAMEO (Figure III.11, left), the emergency managers would navigate through virtual environments (Figure III.11, right).

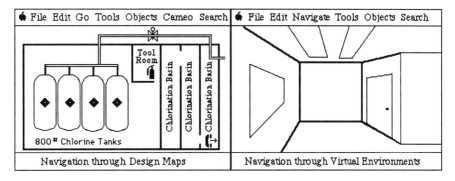

Figure III.11: Hypermedia navigation (left) and virtual world navigation (right).

These virtual environments give a better comprehension of the facility one is walking through. This is another example of architectural walkthrough. However, emergency managers working with a hypermedia system would have the capability to "jump" to any other piece of the facility or the environment.

Adding 2D monoscopic or 3D stereoscopic vision and 2D or even 3D sound has the potential to improve significantly the user's perception of the situation. Emergency managers who have to enter a burning or contaminated building could first analyze the lay-out and the escape routes of the building in cyberspace.

For most tasks, a slide view of the virtual environment might be sufficient. However, some emergency management situations could call for continuous real-time simulation. Commercial software systems are available for continuous VR simulation. An example is the Windows application WorldToolKit from SENSE 8 Corporation, which has a library of over 400 functions to create real-time interactive 3D simulations. The user can create stand alone applications that use Excel or Lotus files.

B) Tele-Virtual Conferencing in Cyberspace

A fundamental philosophical question is: what are the space-time coordinates of cyberspace and its actors? An answer to the first part this question, regarding location in cyberspace, would go beyond the scope of this text. The latter part, however, the coordinates of the actors, is easier to address. Advances in information and communications technology make world-wide real-time communications possible. Information highway systems, satellite communications systems, and increasing computing power allow remote, mobile actors to join a communication meeting at any time in cyberspace.

Commercial teleconferencing systems provide each participant with a view of all the other participants in different windows. Despite the availability of the teleconferencing environment, however, real world meetings still seem to be preferred. Consequently, only environments that closely emulate real world scenarios would be a satisfactory substitute for face-to-face meetings; an example is tele-virtual conferencing [Ramanathan et al. 1992].

In tele-virtual conferencing, each participant sees the other participants in the same meeting room. To generate the stereoscopic vision of the meeting, every participant is being recorded by two (instead of one) video cameras. This picture is then placed around the meeting table in cyberspace. To see the 3D room on the screen, the participants use 3D glasses.

Tele-virtual conferencing can have a crucial impact on emergency management. The various commanders would not have to be present at the response site, but could meet in cyberspace. In addition, one participant could simultaneously participate in several meetings, just by switching forth and back between different cyberspaces. For example, the commander of the coast guard could meet with the on-site response team and instantaneously switch to his/her on-scene commanders to implement further actions. Furthermore, a commander would not even have to be positioned at a headquarters for a meeting in cyberspace. S/he could be traveling on a plane and working with a notebook with two small cameras attached at the corner of the notebook. The 3D glasses and earphones put the emergency manager, via satellite communications, right on the scene. Figure III.12 shows the concept of mobile tele-virtual conferencing.

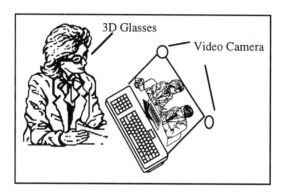

Figure III.12: Mobile tele-virtual conferencing.

C) Training in Cyberspace

The success of an emergency operation depends to a large extent on the training provided to the emergency responders. Today's training consists of exercises with simulations, such as in nuclear power plant drills, fire-fighter training, emergency responder training, training of hazardous materials drivers, and many more. Although the objects, trucks, roads, hydrants, etc. are real, the sensation of a real emergency situation is difficult to create in these simulation "gaming" situations.

Training in cyberspace would not focus on handling devices, shutting down engines, or applying first-aid techniques to people. It would focus on finding ways of getting through a building, finding the locations of hydrants, locating sensitive devices, and finding ways through smoky rooms.

Different scenarios could be devised, focusing on different types of emergency situations, such as emergency responses in contaminated nuclear plants and environments or fire-fighting under dense smoke. While the trainees act under difficult conditions such as a smoke-filled room that restricts visibility and impairs breathing, the trainers could stand next to them in a "clear" cyberspace (i.e., without the artificially generated impacts of the smoke) and control the trainees' actions and advise them. At any time during the simulation, the trainers can change the environment of the cyberspace; for example, they could create more or less smoke, increase the temperature, make a building fall apart, or switch instantaneously from summer to winter conditions.

One training resource is being developed by firefighters in Raleigh, NC, USA and researchers at North Carolina State University's School of Design [Egsegian, et al., 1993]. They are working on creating virtual fires which can be used to train firefighters. Their prototype will model a single-story house fire. The firefighters will respond to the fire with various tactics, and the fire will either be extinguished or reach flashover. During the simulation, the instructor will be able to monitor all aspects of the training scenario, as well as the health of the firefighting trainee.

Another application of VR in training is already being implemented and uses simulators similar to flight simulators. The emergency responder, usually a driver of a vehicle, moves in cyberspace. Vision, sound, and motion are simulated. The trainer can at any time change the conditions of the cyberspace to test or assess his/her behavior. For such purposes, a real vehicle is usually constructed and incorporated into the simulation.

D) Teleoperation and Telepresence

A last major breakthrough in emergency management due to VR is remote operation of robots: teleoperation. The emergency operator's actions in cyberspace are directed to a real robot in the field. Cameras positioned on the robot provide the emergency operator with a 3D view of the real world. Unlike other types of VR discussed in this section, teleoperation uses an environment generated by the real world, rather than a computer-generated environment.

An example of an advanced teleoperation device is a suit that an emergency manager would wear. This suit would transform the emergency manager's actions directly into the robot's actions: the emergency managers and the robot merge in cyberspace. Applications of teleoperation include activities in environments posing physical hazards to human emergency responders, such as sinking ships, earthquake areas, or areas that have been radioactively or chemically contaminated.

In telepresence, the human operator of the remote object receives enough information in a sufficiently natural way that s/he feels physically present at the remote site [Sheridan, 1992]. The degree of telepresence in any given teleoperation depends on the persuasiveness and realism of the teleoperator's cyberspace. This raises the question of what level of realism is appropriate [Zeltzer, 1992]. Because it is not possible to simulate the real world in all of its details and complexities, system designers must decide which sensory cues are most important to perform a given task. For example, in a system designed to train firefighters, the simulation of heat and sound are as important as visual simulation, in order to at least approximate the physical environment that the firefighters will face.

3.4 Conclusions

The significance of virtual reality for emergency management has been recognized. Interactions in cyberspace, teleoperation, navigation in virtual worlds, and video teleconferencing have become viable technologies. However, the role of virtual reality in problem solving and decision making must be addressed in the context of the problem at hand. This chapter introduced some of the current and potential uses of virtual reality in emergency management.

Both the research community and people working in the field have begun to appreciate the potential of virtual reality. New methodologies to improve decision making and problem solving in emergency management will emerge shortly. However, any proposed analytical paradigm must be assessed by gaming or in an experimental setting. Only such a thorough assessment can provide significant enough insights to be useful in the design of virtual reality decision support systems.

CHAPTER IV: ROUTING OF HAZARDOUS MATERIALS

1. The ORM Environment

1.1 Introduction

Safety approaches in hazardous materials transportation are traditionally based on regulations, such as routing restrictions for tunnels, bridges, and curfews. Regulations are difficult to enforce, however, and can be inconsistent among different authorities or countries, resulting in "chaotic, unpredictable" routings that jeopardize rather than promote public safety [AAA, 1986]. The transportation research community started to address hazardous materials transportation as a network optimization problem only about two decades ago. An overview of strategic routing of hazardous materials is given in [Turnquist and Zografos, 1991].

There are other reasons for guiding hazardous materials vehicles during the shipments besides strategic planning and regulations. One reason is the significant increase in complexity in hazardous materials transportation. Changes in Europe, such as openings of borders within the EU and to the East, have tremendous impacts on hazardous materials shipments. Daily operations will be confronted with unexpected situations (e.g., different languages, different safety standards among countries, and new demands); and real-time guidance may become indispensable.

Another reason for real-time guidance is the lack of resources to conduct extensive data collections for strategic route selection. For example, local officials in Virginia decided to prepare for accidents involving only the most dangerous materials, rather than tracking and updating all hazardous materials traffic as a basis for strategic route selection [AAA, 1986].

A third reason is the ongoing change in dispatching that currently takes place due to the commercialization of advanced tracking and communications systems. Commercially available tracking systems can display on a map-like screen the locations of vehicles in real-time, and two-way communications enable data exchange between the dispatcher and the drivers. In fact, Inmarsat and RAM Mobile Data announced a partnership in May of 1997 to provide Inmarsat-C satellite links for the United Kingdom transport and distribution market. The objective is to provide mobile communications in remote areas that are not covered by terrestrial wireless networks. The partnership is expected to be extended to more European countries; this extension should be especially helpful to the trucking industry.

There are also several government efforts that promote real-time tracking. Examples are the U.S. Global Positioning System (GPS), the DRIVE project by the EC, and the EutelTracs system in Europe. Other efforts include several European projects, such as Prometheus or ERTIS (a program to develop systems that automatically communicate motor freight information such as truck locations, speed, road conditions, fuel consumption, and destinations to headquarters) [Jurgen, 1991].

The earliest studies of real-time decision support models for rail and road transportation systems include a pacing model for rail [Kraay et al., 1991], a dispatching model for trucks based on dynamic programming [Powell, 1990], and an integer programming model for dispatching petroleum tank trucks [Brown and Graves, 1981]. Schrijver and Sol [1992] addressed the issue of planning support for fleet management of road transportation, and Sheffi [1991] outlined a shipment information center that provides decision support for transportation logistics including pre-shipment, in-transit, and post-shipment.

These examples of decision support for real-time tracking systems mainly address economic aspects (time minimization, theft, loss control) rather than safety issues. The model presented in this chapter supports a dispatcher in finding safe and cost-effective routes in a continuously changing real-time environment. Whenever a sudden event occurs that affects safety and/or transportation costs of one or more hazardous materials shipments, the dispatcher must assess the impacts on safety and on transportation costs. New routing alternatives are then automatically computed and presented to the dispatcher.

1.2 Operational Risk Management for Transportation of Hazardous Materials

The real-time environment for shipments of hazardous materials consists of three components:

1) The **human dispatcher** (operational risk manager) monitors the transportation system for real-time events (RTEs) and guides vehicles during RTEs. The guidance of hazardous materials shipments in real-time is based on the premise that the dispatcher can better monitor the environment for RTEs and make better decisions than the drivers. Monitoring is supported by satellite technology, traffic police stations, environment monitoring systems, and commercially available software packages that can download via modem the latest

forecast from private weather services, governmental forecasts, and satellite photos.

2) The **transportation system** consists of transportation network, vehicles, surroundings (e.g., urban, rural, and environmental areas), and RTEs (e.g., snow storms, road accidents, vehicle failure).

3) The **communications links** between headquarters and the transportation system can transfer any kind of data (e.g., location of vehicles, condition of vehicles and cargo, as voice, text, graphics). Figure IV.1 illustrates the principle of real-time management of hazardous materials shipments.

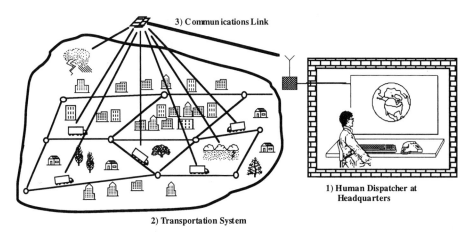

Figure IV.1: The real-time environment for hazardous materials shipments.

The tasks of the dispatcher are the following:

1) **Monitoring** of the transportation for RTEs. An RTE is said to be perceived when the dispatcher considers it to have an impact on safety and /or transportation costs. As soon enough information to determine the area which is affected by the RTE (e.g., area of a snow storm or exact location of traffic accident) has been gathered, the RTE is said to be located.

2) **Identification** of vehicles affected by the RTE. As soon as an RTE is located, the vehicles that plan to drive through the affected area or those already within that area are said to be affected by the RTE. An

algorithm checks which vehicles are affected by the RTE and at approximately what time those vehicles will reach the affected area.

3) **Assessment** of impacts of RTEs on safety and transportation costs for the affected area. The principle of assessing the impacts of RTEs is closely related to the rerouting principles and is described in the next chapter. When the impacts on risk and costs are assessed by the dispatcher, the RTE is said to be assessed.

4) **Rerouting** of affected vehicles. After the areas affected by the RTE have been assessed, an algorithm determines new optimal routes for the affected vehicles from their locations to their planned destinations. Rerouting suggestions can be, (a) to stay on the planned route because the RTE is not that severe, (b) to reroute to a safer or/and more cost-effective route in order to avoid the RTE, or (c) to stop at the next node.

RTEs are sudden, often unexpectedly occurring events that threaten the successful completion of the planned routes of hazardous shipments. The RTEs implemented into the prototype decision support systems were RTEs which had occurred and were reported in news accounts. Every RTE was documented by a short description and a photograph. It must be emphasized, however, that RTEs are events that have the potential to increase risks and costs. It is our contention that undesirable events occur on a relatively frequent basis, although they might not result in reroutings.

The RTEs can be divided into four classes: single and multiple point and area events. A **point RTE** (PRTE) is an RTE that affects only one road link, such as a road accident on a highway segment. One possible route would be to exit and reenter the highway, bypassing the affected highway segment, where a highway segment is defined as the road section between two consecutive entrance/exit ramps. The optimal route to the destination, however, might not be to take a local highway to avoid the affected highway segment.

An **area RTE** (ARTE) affects multiple links either partially or completely. Examples of area events are large weather disturbances like snow storms. The difference between an ARTE and a PRTE is that with an ARTE, the dispatcher assesses the impact of the event on multiple links. This has two important consequences. First, a detour could still lead through the affected area, possibly on an alternate route. Thus, it is not merely an avoid vs. maintain-planned-route decision, like for PRTEs. Second, an ARTE can be assessed as a PRTE by predefining how a single area assessment (for risks and costs) would be distributed among road segments of different classes (e.g., highways and mountain roads). For example, if a snow storm increases the travel risk to a certain level, this new risk value must be interpreted differently for highways than for local roads. In fact, this capability of "interpreting" point assessments for area events was incorporated in the prototype decision support systems for the OP and MAU models. The advantage is that it reduces the assessment time significantly - a crucial aspect in real-time decision making. However, when the dispatcher is presented both the planned and the new routes, s/he can still decide to evaluate the area-wide assessment on a link-by-link basis if the new routes are not satisfactory.

The decision problem is whether to keep the planned route or whether to suggest a new route for the vehicles affected by the RTE from their current position to their planned destination. The accumulation of risks is not considered, and consequently issues of risk equity, in the case of several vehicles using the same routes, are also not addressed. However, the dispatcher is allowed to assess the impact of any RTE on a scale from "no-impact" to "close down affected area." Thus, an RTE does not necessarily have to mean that the affected link or area must be avoided. Before proceeding with the assessment of the RTEs, some propositions are stated which support decision making by attribute over decision making by alternative; that is, they support the need for assessing the impacts of RTEs – preferably by multiple experts.

- If the planned route passes through an ARTE the reroute might still go through (parts of) the area affected by the ARTE. Let R be the planned route and R^* the suggested reroute through the affected area. Assume there is a reroute (R^{**}) which avoids most but not all of the affected area. This reroute can certainly be less risky or costly than the reroute which avoids the whole area. Let l_{ij} be the link between node i and node j, R (the planned route)=$\{l_{12},l_{23}\}$, $\{l_{12},l_{23},l_{45}\} \in$ ARTE, $R^*=\{l_{14},l_{43}\}$ the reroute that avoids the ARTE, and $R^{**}=\{l_{14},l_{45},l_{53}\}$, $\pi(R)=\omega$, $\pi(l_{14})$=LR, $\pi(l_{45})$=LR, $\pi(l_{53})=\omega$, $\pi(l_{43})$=3LR, $\pi(l_{12}|\text{ARTE})$=3LR, $\pi(l_{23}|\text{ARTE})$=2LR. Then, $\pi(R|\text{ARTE})$=5LR is less desirable than $\pi(R^*)$=4LR which is less desirable than $\pi(R^{**}|\text{ARTE})=2\text{LR}$. Consequently, searching for routes that avoid ARTEs might lead to inferior solutions.

- If n individually occurring PRTEs all cause reroutes (i.e., the assessments of risks and costs suggest alternate routes), then the best reroute when all n PRTEs (with the same assessments) occur simultaneously might cross some of the RTEs. Let l_{ij} be the link between node i and node j, R (the planned route)=$\{l_{12}, l_{23}, l_{34}\}$, $PRTE_1 \in l_{12}$, $PRTE_2 \in l_{34}$, R_1^* (the reroute that avoids the link affected by $PRTE_1$; i.e., l_{12})=$\{l_{13}, l_{34}\}$ and $R_2^* = \{l_{12}, l_{24}\}$. Then, with l_{23} being a one-way link, $R_{12}^* \in \{R_1^*, R_2^*\}$ is the best reroute if neither PRTE shuts down the affected link. However, if the affected link is shut down, no reroute at all might be possible. This result implies that searching for a reroute by avoiding all RTEs (given that one-by-one they should be avoided) does not necessarily lead to the best reroute. Consequently, decision making by attribute should be employed; i.e., the computation of the best reroute should be based on an assessment of the impacts of the PRTEs.

- The reroute of n simultaneous PRTEs, where at least one PRTE does not affect the planned route (R), is not necessarily identical to the reroute of the simultaneously occurring PRTEs, all of which affect the planned route ($PRTE_i \in R$). Let $R_1 = \{l_{12}\}$, $PRTE_1 \in l_{12}$, $R_1^* = \{l_{13}, l_{32}\}$. Then, $PRTE_2 \in l_{13}$ contradicts the assumption of the negation of this preposition. This result implies that searching for a reroute by avoiding PRTEs which affect the planned route can lead one to other RTEs which do not directly affect the planned route. Consequently, decision making by attribute should be employed.

- The minimum time available to make a reroute decision is the time it takes a vehicle to reach the first decision point; i.e., the end of the current road link, such as the first exit from the highway. The maximum decision time is the time it takes a shipment to reach the last decision point; that is, the last node where a decision can be made to avoid the affected area. Decisions made within the minimum decision time are optimal, while those made within the maximum decision time are at least feasible. The maximum decision time can be computed by shutting down the affected area and calculating the travel time to the reroute point; that is, decision making by attribute.

Three additional management aspects are as follows.

1) **Strategic and Real-Time Assessment**: Assessment of some "known" RTEs can be done strategically. These are the RTEs that occur relatively often and therefore are fairly well known. Examples are annually occurring snow storms in the mountains or typical congestion and accidents on highways. The less expected and less well known an RTE is (i.e., the more exceptional the RTE), the less strategic support

can be provided to the dispatcher for real-time assessment of risks and costs. Under normal conditions, however, there will be very few HR entities because regulations restrict the transportation network to rather safe links. This implies that under normal conditions, shipments of hazardous materials primarily follow a cost-driven route selection, while during RTEs, risk-driven route selection will usually be the case.

2) **Time Frame for Decision Making**: A time frame for risk assessment and decision making during an RTE can be determined for each vehicle (see Figure IV.4). The minimum decision time, T_{min}, is the time it takes the vehicle to reach the end of the link it is on when the RTE happens, because shipments can change their planned route only at the nodes. The maximum decision time, T_{max}, is based on the worst-case assumption that all links affected by the RTE must be avoided. This means that T_{max} is the time it takes the vehicle to reach the last node on the planned route, from which a route to the planned destination exists that avoids the affected region. Guidance decisions made before T_{min} will be optimal, and those made before T_{max} are certainly feasible. Decisions made after T_{max} might be not feasible if the RTE forces the shipment to avoid all affected links (a preference).

3) **Management of Affected Vehicles**: RTEs occur and terminate randomly and shipments of hazardous materials start and end at different times of day. Therefore, new optimal routes must be computed at the beginning of an RTE for all affected vehicles, and after termination of an RTE for all affected vehicles plus for those that started during the RTE.

As an example, let RTE_i be the i-th RTE and $S_j\{.\}$ the j-th shipment affected by a set $\{.\}$ of RTE_i's. Let's assume that the RTEs occurred and the shipments started in the following sequence: $S_1\{1,2\}$, $S_2\{1,3\}$, RTE_1, $S_3\{-\}$, RTE_2, $S_4\{3\}$, RTE_3. According to this sequence, S_1 and S_2 had assigned their optimal routes under normal conditions, since no RTE was going on when they started their trip. However, their routes have been reassessed twice, once for each RTE that occurred while they were on the road (S_1 for RTE_1 and RTE_2, and S_2 for RTE_1 and RTE_3). Although S_3 is not affected by any RTE, it might not be on the most desirable route under normal conditions, since the computed route considered that RTE_1 was present. For S_4 the situation is similar to S_3, except that it is also affected by RTE_3. When RTE_2 is over, then S_1 and S_4 must be reevaluated. S_1 must be reevaluated because it was affected by RTE_2, and S_4 because it started the shipment after RTE_2, although it was not affected by this RTE. This example indicates the complexity of real-time management. It is thus obvious that computerized decision support is indispensable and that it must be tailored to the cognitive capabilities of the human dispatcher.

2. ORM Decision Logic for Transportation of Hazardous Materials

2.1 General Aspects

Decision models that recommend routes should support the dispatcher in: (i) the identification of the affected area on the network, (ii) the assessment of the impact, (iii) the identification of the affected vehicles, and (iv) the rerouting decisions for the affected shipments. The assessment of the impact of an unforeseen RTE should be done by altering the risk and cost values (as well as the values of any other attributes used in the routing model) for traveling through the affected area.

The optimal routes for both the ordinal preference (OP) and multiattribute utility (MAU) models are based on the shortest-path principle with either a quasi-linear preference function (OP model) or a linear preference function (MAU model). If a shipment is affected by an RTE, new optimal routes are computed from the shipments current position to the planned destination. Despite the limitations of using shortest-path based algorithms to reroute vehicles (or even a truck fleet), especially when risk is a criterion, they are the most frequently used routing algorithms for in-vehicle real-time routing [ITS, 1995].

The assumption of the linear model for a shortest-path based model for hazardous materials transportation (HMT) routing can be questioned if the routing criterion is probabilistic risk. Boffey and Karkazis [1995] have addressed the problem of linear vs. non-linear minimum risk routing models and derived a condition which can ensure that both linear and non-linear models result in the same origin-destination route. The frequently used additive HMT routing model computes the risk of the total route from origin to destination as the sum of the link risks, where the link risk is the expected link damage. The multiplicative model computes the route risk as the expected damage associated with the whole route from origin to destination; that is, the sum of conditional expected link-damages, where the condition is that no accident has occurred on previous links.

The additive model is in fact an upper bound for the multiplicative model [Boffey and Karkazis, 1995].

2.2 Ordinal Preference Routing Model

The OP model is based on a lexicographic preference structure, with the classes: link closed (α) $\prec$ high risk (HR) $\prec$ transportation costs (C) $\prec$ low risks (LR) $\prec$ negligible impact (ω). The preference of a link (l_{rs}) between node r and s is defined as: $\pi(l_{rs}) = \sum_{i=1}^{5} M_i v_{irs}$, where the v_{irs} are the values from the classes HR, C, or LR, and the M_i are large negative numbers that discriminate $\sum_i v_i$ such that they remain incommensurable: $M_{LR} \gg M_C \gg M_{HR} \gg M_\alpha$, and $M_\omega = 0$. The routing model for the OP system, for a route between *start* and *end*, is then ($M_\omega \equiv M_1$, $M_{LR} \equiv M_2$, $M_C \equiv M_3$, $M_{HR} \equiv M_4$, $M_\alpha \equiv M_5$), where $x_{rs}=1$ if link l_{rs} is part of the route, and $x_{rs}=0$ otherwise:

maximize: $\sum_{r,s \in I_N} x_{rs} \sum_{i=1}^{5} M_i v_{irs}$

subject to:

- $\alpha \prec v_{irs}$, $\forall i,r,s$

- $\sum_{i=1}^{n} x_{ik} - \sum_{j=1}^{n} x_{kj} = \begin{cases} 1, & \text{for } k = start \\ 0, & \text{otherwise} \\ -1, & \text{for } k = end \end{cases}$

- $x_{rs} \in \{1,0\}$

The routing algorithm, as implemented into the OP and MAU models, did not use the "big M method" [Hillier and Lieberman, 1986], but used logic statements to circumvent the possibility of incorrect computations if the number of risk sources (or costs) becomes too large.

2.3 Multiattribute Utility Routing Model

The MAU model uses a cardinal scale for both risks and costs. The risk values, $x \in [0,100]$, where 0 stands for no risk and 100 for certain accident, are transformed into absolute values, $\lambda(x) \in [0,1]$, with the transformation proposed in Chapter II, Section 2.4:

$$\lambda(x) = \begin{cases} \bullet \ 10^{-[10-INT(x/10)]}, & \text{for } x \in Z = \{0,10,20,\ldots,70,80,100\} \\ \bullet \ \left[x - 10 \times INT\left(\dfrac{x}{10}\right)\right] \times 10^{-[10-INT(x/10)]}, & \text{for } x \in [0,100] \setminus Z^* \end{cases}$$

For non-highway links, the risk values are appropriately transformed to reflect the corresponding road type (e.g., canton or mountain road). The absolute risk values are then transformed into risk-costs using estimates for life-saving costs (LSC) of 10^7 Swiss Francs (*SFr*), which were confirmed by recent research [Glickman and Sontag, 1995]. The transportation costs are computed using the estimate of 1.0 *SFr/km* for driving at a velocity of 80 *km/h*. Thus, the transportation costs, c, along a link of length l at a velocity v (*km/h*) are $(80/v) \times l$. The total costs are therefore $C = c + 10^7 \times \lambda$. For example, if a truck drives 250 kilometers at a velocity of 100 km/h and the risk is estimated to be $x=43$, then the total costs are $C = (80/100) \times 250 + 10^7 \times 3 \times 10^{-6} = 200 + 30 = 230$ *SFr*.

2.4 Decision Making

The route-selection algorithm is based on the shortest-path algorithm [Hillier and Lieberman, 1986]. For the OP Model, the principles of the preference algebra are used as proposed in Chapter III. The algorithm chooses the optimal route (considering the two criteria 'risk' and 'cost') from the origin of the vehicle to its planned destination by combining links according to the following priorities: (1) it never takes a 'closed' link, (2) among the 'open' links, it avoids as much as possible the high-risk links, (3) if there are no high-risk links (or for ties), it takes the most economical route, and (4) if there are still ties, it avoids as much as possible the low-risk links.

The routing problem can be solved with a network search algorithm similar to the shortest path algorithm. It fans out from the origin and determines continuously the most desirable route to all the other nodes until the destination is reached. It can be summarized in the following way: assume that at some stage k in the computing process the preferences and the paths of the most desirable routes from the origin O to the nodes N_i's are known, where $N_i \in N = \{N_1,\ldots,N_k\}$. The nodes in N are called closed nodes. Let $M_j \in M$ (set of adjacent nodes of N, $M_j \notin N$) with the preference $\pi_{N_iM_j}$ to go from N_i to M_j. The algorithm is the following:

(1) <u>Initialize</u>: Choose (close) the origin node O.

(2) <u>Close Node</u>: Choose and close (the preference to the origin is definitive) as the $(k+1)$-th node, M_j, where j satisfies $\alpha \prec \pi_{N_iM_j}$ and $\pi_{ON_i} \oplus \pi_{N_iM_j} \sim \text{Max}(\pi_{ON_i} \oplus \pi_{N_iM_r})$, where $i=1,2,...j,...,k$, $M_r \in M$. Ties are broken arbitrarily. In words: M_j is a direct successor of an already closed node N_i. In addition, it has the highest preference to the origin among all the direct successors of all the closed nodes. Thereby, the preference of link-exceeding entities is counted only once.

(3) <u>Loop</u>: If the destination D is not in N, then go to (2), i.e., if M_j D. The preference to go from the origin to M_j is $\pi_{ON_i} \oplus \pi_{N_iM_j}$ and the most desirable path is via N_i.

The optimal route for the MAU model uses the total cost values (i.e., the sum of weighted risk-costs and transport costs), and computes the minimum cost route as follows (with n nodes on the network, e.g., intersections; and $x_{ij}=1$ if the route includes the segment between node i and node j, and $x_{ij}=0$ otherwise):

- of all the possible routes between the vehicle's position (O) and its planned destination (D):

$$\sum_{i=1}^{n} x_{ik} - \sum_{j=1}^{n} x_{kj} = \begin{cases} 1, & \text{for } k=O \\ \text{otherwise} \\ -1, & \text{for } k=D \end{cases}$$

- choose the one which minimizes costs: $\sum_{i=1}^{n}\sum_{j=1}^{n} x_{ij}\left(\dfrac{80}{v_{ij}}l_{ij} + LSC \times \lambda_{ij}\right)$

The number of comparisons needed to determine the next node to be closed for both the OP and MAU models in step 2 of the algorithm is k (comparing $k+1$ points).

Thus, in an n-node network an upper bound for the number of comparisons is: $1+2+...+(n-1)=n(n-1)/2$. This number could be reduced by fanning out from origin and destination simultaneously.

Both the OP and MAU models are based on the principle of the shortest path algorithm, where the resulting routes are presented to the dispatcher for decision making. However, shortest-path like approaches should not provide only one best route but several options for the decision maker to choose from because of the risk criteria [Boffey and Karkazis, 1995] and of the multiattribute structure [McCord and Leu, 1995]. Both the OP and MAU not only give the dispatcher the reroute and planned route for each vehicle, but they also provide the capability to select a completely different route simply by clicking with the computer mouse on the appropriate links on the road network.

For the transportation network used for the OP and MAU models the safest route under normal condition is always the most economic route; a result that may not be valid if risk (and thus route preference) is based solely on human lethality [Glickman and Sontag, 1995]. Thus, the additive approach is appropriate.

Both the OP and MAU models for real-time impact assessment and rerouting used in the experimental assessments (Chapter VII) are based on risks and costs to determine the most desirable routes for the shipments. The transportation network which is being monitored by a dispatcher has been assessed link-by-link for risks and costs for normal driving conditions. The dispatcher then can at any time change the risk and cost values of the affected links and the system will compute alternate routes for all the affected shipments.

2.5 Example of Preference Routing Model

An example of the algorithm is given here, where the entities are the links and the light gray areas are assessed entities on the network (see Figure IV.2). During normal conditions, no HR entities are present. The assessment of the entities under normal conditions is done strategically for the entire transportation network.

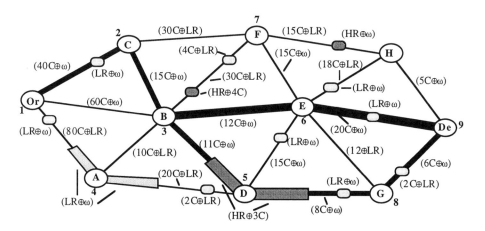

Figure IV.2: Example of routing algorithm.

When no RTEs are present, the optimal route computed by the preference logic is determined solely on the basis of cost, since HR entities are not present. The numbers next to each node indicate in which sequence the nodes have been closed. The thick line shows the most desirable path from origin (Or) to destination (De) under normal conditions. The node sequence is Or, C, B, D, G, De. The overall preference is 82C⊕3LR. An RTE affects the dark gray areas which under normal conditions had neither risk nor cost concerns (ω preference). It is assumed that the dispatcher assigns a HR risk preference to each of these areas. The most desirable route during the RTE has the node sequence Or, C, B, E, De, and the overall preference is 87C⊕2LR.

3. Decision Support Systems for Transportation of Hazardous Materials

3.1 Capital District of New York State, USA
A) Components of the Decision Support System

A transportation network from the Capital District region of New York State has been chosen that consists of 69 links. To allow for various transportation environments, a selection of highways, rural roads, and inner-city roads have been considered. To include longer distances, the map has been distorted so that all links do not necessarily have the same scale. To compensate for this and make the simulation realistic, the vehicles have been assigned a lower speed on links that are actually longer than they appear on the map.

Figure IV.3 shows the Monitoring Screen with the road network on which three vehicles move. The vehicles are either on the road or at a place at the top right of the screen reserved for when they are not on the road (parking-space). Information about the vehicles can be obtained by clicking on them. If a vehicle is on the road, its parking-space shows either an "OK" or a blinking "RR." The "OK" occurs when the vehicle's route is based on all ongoing real-time events. The blinking "RR" occurs when the vehicle's route has to be updated either due to a new RTE or due to the deletion of an RTE. Information about a link can be obtained by clicking on it. The actual time and date are displayed at the right bottom of the screen.

By clicking on the buttons on the right of the map, the user can activate different menus, such as vehicle initialization, monitoring, RTEs definition and deletion, risk and cost assessment, routing, and information transfer to the vehicles. The monitoring or the movement of vehicles can be activated or deactivated by clicking on the button "Monitor." Before any other tasks can be done, the monitoring process must be interrupted. The button underneath the monitoring button allows the user to define or delete real-time events or to obtain information about them. If the field above the button "Not-Assessed RTEs" is "Yes" and blinking, then an RTE has been located (region and name are defined and affected vehicles determined) or deleted, but the affected links have not yet been assessed. The field underneath the button "Assessed RTEs" shows "Yes" if there are any assessed RTEs and "No" otherwise. The next button underneath is clicked when a vehicle must be routed or rerouted. The "Dispatch" button is clicked when a new route must be transferred to a vehicle.

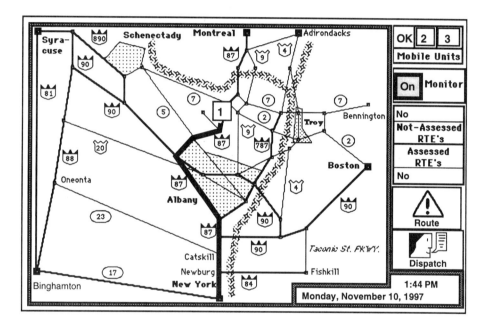

Figure IV.3: The monitoring screen.

Each vehicle has a card in HyperCard that contains information about the vehicle is held. This information includes start time, origin, destination, cargo, planned route, and the real-time event affecting the vehicle's route.

A new real-time event (NRTE) is a real-time event for which the links have not yet been assessed (i.e., the event is located but not yet assessed). After the vehicle has been initialized, the "optimal" route can be determined, but only if there aren't any NRTEs. This is because an NRTE indicates that risk or cost values of some links might change and a reassessment of the routes must be done. The new vehicle should wait until all NRTEs are assessed in order to avoid being rerouted soon after it begins its trip. If no NRTE is present, the vehicle can be routed.

If the vehicle is on the road, the operator can choose between the two menus: "Show Info" and "Re-Route." Rerouting can also only be done when no NRTEs are present. Routing (or re-routing, which is the same process) can have different motivations. The first one comes from the operator and is consequently called "voluntary." This refers to routing a new vehicle that has been idle or to checking alternate routes, eventually on a subnetwork. The second motivation is due to the definition of a new or the deletion of an ongoing real-time event and is called "imposed." No matter what the motivation is, the process of determining a route is the same in all the cases.

The movement or the continuous location of the vehicles on the transportation network is simulated. It is performed by the message handler, which is called the "mover" in this context. When the mover is activated, it processes each vehicle in a serial mode; that is, it moves the vehicles one by one. The delay between each vehicle's processing depends on the mover's speed, which can also be altered. This is especially important when running the program on computers with different processing frequencies.

When other tasks must be performed by the message handler, such as routing or defining a new real-time event, the mover must be set to idle. When reactivating the mover, the operator has the choice between two menus: "New-Start" and "Re-Start." The first choice disregards the time elapsed while the mover was idle, while the second choice takes this time into account. With the "Re-Start" menu, the mover "jumps" the vehicles to the location they would have been in if the mover had not been deactivated for a certain time; i.e., it simulates a parallel processing of multiple handlers. This is especially interesting if time constraints must be simulated for the assessment phase and other processes. While the operator performs tasks other than monitoring, the vehicles move along their routes.

The speed of the simulated vehicle movement depends on a "delayer" called TC (transit control). While the mover wants to move a vehicle at each processing step for one-tenth of the distance between two nodes, the delayer tells the mover to move the vehicle only each $(TC+1)$-th processing step for one-tenth of the distance. Therefore, the higher the TC, the slower the vehicle moves. The move counter is called K and has value 1 at the beginning of a link and value 10 when the vehicle reaches the end of a link. Therefore, a vehicle on a link with a delayer TC of 3 takes 40 processing steps by the mover to travel through the link.

The "jump" done by "Re-Start" is computed for two different situations. The first situation is when the vehicle is still on the same link it was on at the time monitoring was set to idle. In this case, the "jump" occurs by replacing K with $K+dt$, where

$$dt = \text{round}(dt'/(TC+1)), \text{ and } dt' = \text{round}(\text{elapsed time}/10).$$

The reason for dividing the elapsed time by 10 is that the "jump" should correspond approximately to the distance the vehicle would travel in the elapsed time and depends also on the computer's processor.

The second situation in which the "jump" is considered is when the delay was long enough for the vehicle to move past the end of the link it was on at the time the mover was set to idle (i.e., $(K+dt)>10$). In this case, the "jump" is determined in the following way: delete the $INT((K+dt)/10))$ first links of the route; replace K with $(K+dt)-10 \text{ trunc}((K+dt)/10)$.

B) Operational Risk Assessment for Transportation of Hazardous Materials

RTEs can be at three different levels, depending upon what the operator knows of the situation. The lowest level of an RTE is when the operator knows that the RTE is present but cannot determine the affected region (perceived RTE). An example is a major weather change that has not yet been located. An RTE in this initial stage will be recorded by the operator, and if the RTE is of very serious concern, the operator will warn all vehicles.

If sufficient information is available to identify the affected region, the RTE is called "located" or "new" (NRTE). Only at this stage does it become subject to analysis, since the affected links and the affected vehicles can be determined. A vehicle is affected by an RTE if at least one link of its planned route falls into the affected region. If sufficient information is available to assess the new costs and risks of the affected links, the RTE is said to be assessed, and the assessed RTE is abbreviated simply as RTE, rather than NRTE.

Each NRTE and RTE has its own card in HyperCard, on which information about the event is held. The operator can view, retrieve, change, or delete information about any event. Two pop-down menus on the monitoring card in HyperCard deal with real-time events, one with NRTEs the and other with RTEs. Above the NRTE button and below the RTE button is a field indicating "No" if there is no event present or "Yes" if there is at least one event present.

As long as an event is an NRTE, the operator is kept on alert. S/he must gather more information to assess the affected links. Therefore, if there are any NRTEs, the field above the button for NRTEs shows a blinking "Yes;" otherwise, it shows a non-blinking "No." Another alert refers to the affected vehicles. As soon as a vehicle is determined to be affected by an event, its "parking-place" blinks with "RR," which stands for reroute; otherwise, the "parking-place" shows a non-blinking "OK."

A real-time event always enters the DSS as an NRTE, and if sufficient information is available to assess the affected links, it is transformed to an RTE. If the assessment is postponed, the operator is alerted. When the operator clicks on the NRTE button, a pop-down menu appears, which shows the menus "New" and the names of the NRTEs already defined, if there are any. The operator is then asked if the affected region has been marked using the painting tool and the mouse. So, s/he is asked to enter the name of the NRTE. After this has been done, the message handler automatically puts the name, graph, and time into the appropriate information card in HyperCard. The affected links and vehicles are then determined. The affected vehicles have a blinking "RR" in their "parking-place." The system computes T_{min} (time to reach the end of the link it is on right now) and T_{max} (time to reach the first link of the affected region) for each affected vehicle and puts this information into the NRTE information card in HyperCard. The units of T_{min} and T_{max} are not seconds, minutes, or hours but rather the number of jumps simulating the movement of the vehicles. Using the same notations as introduced for the vehicle movement, the two times are defined in the following way:

$T_{min} = TC'+(10-k-1)((TC+1)$, where TC' is actual TC-value $\in [0,TC]$ and
$T_{max} = 10[((TC+1))-(\text{first } TC+1)]+T_{min}$, where goes from 1-st link to last link before 1-st link in affected region.

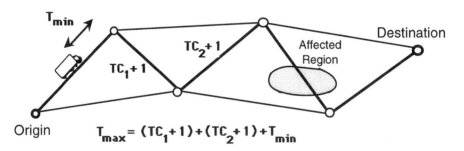

Figure IV.4: Minimal and maximal decision time.

It takes the vehicle T_{min} jumps to reach the end of the link, and it takes the vehicle T_{max} jumps to reach the first link that lies (at least partly) within the affected region. TC_1 and TC_2 are the link-specific delayers for the mover. If $T_{min}=T_{max}=0$, then the vehicle is already within the affected region. In this case, the handler alerts the operator.

After the RTE has been located, the message handler asks whether the affected links can be assessed; i.e., whether the NRTE can be transformed to an

RTE. If not, the case, the message handler exits; that is, it becomes idle and can be reactivated by clicking on another button. If the operator assesses the affected links, the NRTE is transformed to an RTE. This means that the event's name is transferred from the NRTE list to the RTE list. The operator is then automatically presented the list of the affected links for assessment. Once all the links have been assessed, the message handler becomes idle. The next step is to recompute routes for the affected vehicles. If the links have not yet been assessed, the pop-down menu of the NRTE button shows the names of all the NRTEs. If one of these names is selected, the operator has the choice of either assessing the links or deleting the NRTE. If an NRTE is deleted, the information card is deleted, and if no other NRTEs are present, the blinking "Yes" turns into a non-blinking "No."

The menu "Assessment" is for assessing the links whose risk and cost values might change either due to the occurrence of a new real-time event or because an ongoing real-time event is over. The menu can be chosen in different instances. After a new real-time event has been located, the handler asks the operator if the operator also wants to assess the affected links. If this is not the case, the assessment of the affected links is postponed. Selecting NRTE from the pop-down menu, the operator is asked if s/he wants information about the NRTE or if s/he wants to assess the affected links. If the operator does not feel ready to assess the links, s/he can request information about the NRTE. A button called "Assessment" will blink on the information card as long as this NRTE has not been assessed.

Whenever the operator selects "Assessment," the handler presents the operator with a card showing all the affected links. By clicking the appropriate button, the assessment process is initialized. This means that each link's card in HyperCard is presented to the operator (see Figure IV.5). The card tells the operator information about the link: name, length, coordinates, etc. A map shows where the link is located.

Preference assessment considers humans' limitations in risk and cost assessment in an operational environment, where information is usually incomplete and inaccurate. The preference assessment spectrum consists of four risk-preference classes, one cost-preference class, and two preference classes for both risks and costs. The latter two preference classes either close a link because it is too dangerous or too expensive, or they neglect cost or risk aspects. Costs are changed simply by entering the appropriate dollar amount. The risk-preference value for road segments that depend strongly on the segment length is assumed to be identical with the segment length (in miles). The preference value for entities that are not length-dependent is entered by clicking the button "Points." Once the assessment for one link has been completed, the handler presents the operator with the next link, and so on, until all the affected links have been assessed.

An example is given in Figure IV.5. The link is affected by the two RTEs 'Snow' and 'Accident.' The operator increases the operational costs by $1,000 to account for the expected delay due to these two RTEs. The three miles through the urban area are assessed as high-risk and the other 143 miles are assessed as low-risk. One low-risk value is assigned to the one bridge and the one intersection on the link.

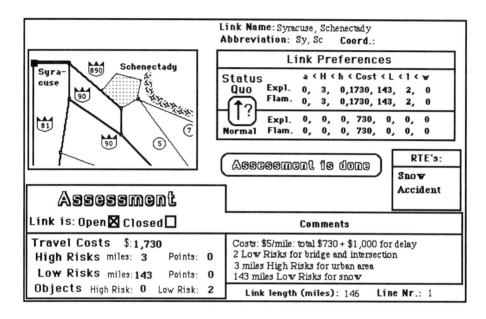

Figure IV.5: Preference assessment card.

C) Route Selection Process

The process of determining a route is performed on a special card in HyperCard. The operator can reduce the network to a subnetwork by using the mouse to select the appropriate region or links. The system then automatically determines the links that belong to this subnetwork.

For demonstration purposes, the routing algorithm is written in Hypertalk. However, to improve execution speed, the algorithm could be written in Hyperprolog, C, or Pascal.

Once the node-sequence of the optimal route from origin to destination has been computed, the handler puts the solution into the appropriate vehicle's information card. If the path-finding process was used for a rerouting, the operator is presented with both graphical representations of the old and the newly computed routes, as well as the overall risk and cost values of the routes. The operator can then select which route s/he wants to suggest to the driver. If the path-finding process was used to determine a new route and the operator does not like the computed route, s/he can either alter the preferences for certain links and/or choose another subnetwork and then reroute the vehicle.

D) Assessment of the DSS

The practical assessment of the DSS was performed with the help of experienced dispatchers at Chemical Leaman Tank Lines, Inc., at its headquarters in Exton, PA, USA, and at its regional dispatch center in Albany, NY, USA. In the first phase, the system as currently designed was introduced to the persons participating in the assessment. The basic tasks of monitoring, definition of real-time events, risk assessment, and routing were introduced. This phase aimed at the assessment of the models and the user interface.

In a second phase, a simulated scenario was run to assess the task and decision flow, as well as the ease of use. The occurrence of real-time events was simulated while the three vehicles were on their route. Since some of the vehicles' routes went through the affected region, and those vehicles had to be rerouted. The simulated run lasted about 20 minutes. In the third phase, the evaluation of both models and DSS were performed by discussing with the operators the design of the DSS and the expected benefits if this DSS were to be integrated in their daily operations.

The initialization of the vehicles corresponds to the dispatching tasks with which the operators are very familiar. The drivers' names, the type of cargo, and origin and destination are assigned to each truck. However, the possibility of monitoring the vehicles on a screen was new to the operators and very much appreciated. The first new task that the DSS provides is support in route selection. The possibility of choosing a subnetwork, to which the vehicle is restricted, gives more flexibility. If many vehicles have to be routed or rerouted, however, repetition of this process might become time consuming and confusing. More automation is appropriate (especially when more than three vehicles must be guided) but has not been introduced in this prototype version of the DSS.

Defining an NRTE made intuitive sense to the operators. The graphical capabilities of the hypermedia environment facilitate crucial tasks, such as selecting affected links, and are much more convenient than searching in a list or on a map for the desired links. The RTE information card in HyperCard provides the important information about the name of the RTE, the time it occurred, the affected links, and especially the affected vehicles and the time available for rerouting them. While operators in traditional settings would need to check a road map to see which vehicles are affected by the RTE, the automatic assessment by the DSS has the advantage of saving time and avoiding errors.

The principles of assessing the affected links in conjunction with the principles for route selection were easily introduced by using the descriptive formulation for route selection given above. The operators readily understood the principles involved but would require more training before they would feel comfortable with the process of preference assessment. With appropriate training, the average operator should be able to use the DSS easily and effectively.

The appropriateness of ordinal measures for risk assessment is supported by government efforts in operational emergency response. Instead of assessing numeric risk values, the emergency responder is more interested in the locations of sensitive objects (hospitals, schools, ground water) and properly equipped response units.

Considering that risk and cost assessment is required only in very specific and infrequent cases, the major tasks of the operator will be monitoring and automatic assessment; i.e., either the RTE is known or the operator assigns to all links the same preference (e.g., close the link). It was evident that too much flexibility can confuse the operator in stress situations and that dull tasks such as monitoring and searching for disruptions should be automated as much as possible. This fact has also been noted by Hopkin [1989], who studied automation capabilities for air traffic control.

The overall assessment of the DSS and its underlying decision models was very positive. Automation certainly has to be emphasized in the practical use of the DSS for a large number of vehicles. For training purposes, however, the DSS as currently designed with three units is appropriate. Only an assessment in a real application will show how much automation is appropriate and how much flexibility an experienced operator needs and wants.

3.2 Switzerland
A) Components of the Decision Support System

A decision support system was constructed to support the transportation of hazardous materials within Switzerland. A road network consisting of 76 links was selected; this network is used for transporting the vast majority of the hazardous material shipments during the winter months.

Both the ordinal preference (OP) model and the multiattribute utility (MAU) model were implemented in this DSS (Figure IV.6). For the MAU model, the risks under normal conditions were computed according to [BAP, 1988;

Glickman, 1991; and Saccomanno and Chan, 1985]. For all six road segments, the risks under normal conditions are in the order of magnitude 10^{-9}. The risk values of each link were transformed to risk-costs by multiplying them with $\$10^7$ (life-saving costs). Hence, for each link, an overall disutility between 0 and 100, consisting of a risk and a cost component, could be computed. It is interesting to note that under normal conditions, the risk component is approximately only 2% of the overall disutility, which means that optimal routes under normal conditions depend solely on transportation costs; that is, the transportation network is safe.

For the OP model, all highway segments of the transportation network were assigned a preference of ω (no risk) under normal conditions. All other road types (urban and rural canton roads, mountain roads, and tunnels) were assigned a preference of LR. Additional entities were road segments with narrow curves, bridges, tunnels, mountain roads, highway segments with very high average daily traffic, road segments through highly populated areas, and road segments through sensitive environmental areas. They all have LR preferences under normal conditions. HR and α preferences could therefore only occur during RTEs. The transportation costs (C) for both the OP and the MAU models were set at $1 per kilometer when driving 80 km/h.

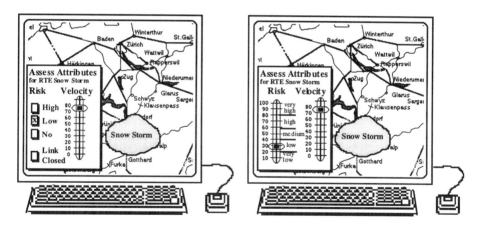

Figure IV.6: Interfaces for assessment of risk and costs for OP and MAU models.

B) Scenario Generation: Vehicles and Events

The major purpose of the hypermedia system is to simulate decision situations in operational vehicle dispatching. These situations are summarized in scenarios. A scenario consists of a combination of vehicles with their planned routes and a set of events that affect the safe and efficient shipments of some of the vehicles. The current default values are three vehicles and two events that occur simultaneously after the vehicles have traveled for a certain time.

The procedure of a scenario depends on the modeling paradigm (see Chapter VII, Figure VII.2). For all approaches, the routes of the vehicles are announced by a voice and displayed on the map. Then, the vehicles start their shipment by driving along their route. These movements are animated on the screen. After some time, RTEs occur. They are announced by a voice and a short text, and displayed on the map. A representative photograph appears on the screen. From then on, the procedure of the scenario changes depending on the decision support paradigms.

The OP and MAU approaches ask the operator to assess the risk and velocity values for the area that is affected by the event. Thereafter, new routes are computed, but the dispatcher (operational risk manager) can always suggest a different route.

Not all RTEs necessarily increase risks and costs. If risks and costs do increase (i.e., hazardous RTE), an RTE is considered to be assessed correctly if the rerouting suggestions avoid the RTE. If risks and costs do not increase (i.e., harmless RTE), an RTE is considered to be assessed correctly if the planned routes are still proposed to be optimal.

C) Robustness of Routes

Every optimal route under normal conditions has a certain robustness compared to the optimal route that avoids the area affected by the RTE. The more risk and/or velocity can be altered without affecting the optimal route, the more robust this route is. Consequently, correct assessment depends on the robustness of the optimal route under normal conditions. Figure IV.7 shows an example for a hazardous RTE and a non-hazardous RTE for the preference model (OP) and the utility model (MAU). The risk/velocity values under normal conditions are also given. The dispatcher can increase risks up to 100 for MAU and up to a for OP, and reduce velocity down to 0 for both models. For the hazardous RTE, the initial risk/velocity pair (under normal conditions) is wrong for both models; dispatchers must move the risk/velocity pair outside the dotted area into the striped area (see Figure IV.7). For the non-hazardous RTE, the risk/velocity pair under normal conditions is correct for both models; subjects must stay within the striped area. Six examples of possible reassessments are given in Figure IV.7: P3 and P4 are correct for both models; P2 is correct for OP but wrong for MAU; P5 is correct for MAU but wrong for OP; P1 and P6 are wrong for both models.

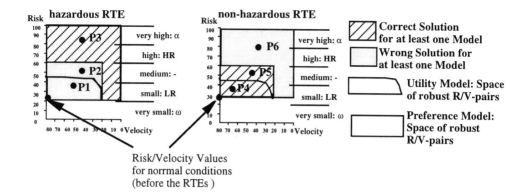

Figure IV.7: Robustness of routes.

3.3 Conclusions

The integration of the decision logic of operational risk management with advanced communications and positioning technologies provides the basis for traffic management in general as well as for the specific case of transporting hazardous materials. When a traffic assignment involving hazardous materials occurs, emergency response operations must be undertaken. The next chapter shows how the operations can be supported by operational risk management.

CHAPTER V: EMERGENCY RESPONSE

1. The ORM Environment

1.1 Introduction

Emergency response typically involves the activation of one or more contingency plans. These plans are implemented by the establishment of a command and control center. The "commander" on-scene coordinates the activities of his or her staff who gather data, do analysis, make decisions, and monitor their implementation. The activities required to respond to an incident are dangerous, and must be performed under time pressure.

Emergency response is based upon an assessment of the potential impact of an accident and the courses of action needed to eliminate or at least mitigate this impact. However, these contingency plans are devised assuming that once they are activated, response activities can be carried out with certainty - an expectation which turned out to be unrealistic in the case of the Exxon Valdez accident [Harrald et al., 1992]. Therefore, a flexible approach to emergency management is needed that deals with uncertainty and allows revisions of planned courses of action and, if necessary, the development of new plans – within an analytical framework that considers the costs, risks, and benefits for response activities.

During the response operation, unanticipated events may arise that affect the planned activities. For example, traffic congestion could delay the arrival of the response team on the scene and bad weather could prevent needed response equipment from arriving on site. The "commander" must be able to assess the potential impact of these events and decide whether to continue the planned course of action (CA) or to redirect it to alternate activities to maintain safety and efficiency of the operation. Therefore, real-time monitoring and control of both response activities and external events that have the potential of affecting these activities must be an integral part of effective emergency response.

Monitoring and controlling the response activities and the potential external events requires gathering and processing of data, such as the extent and severity of the event, meteorological conditions, and current capabilities for response. These data are then assessed, courses of action developed and evaluated, and decisions made in a very time-constrained environment.

The quality of these decisions depends on the quality of the data and the operational risk manager's ability to process these data in a form that is understandable and useful to the emergency response team. In addition, initial decisions act as constraints on subsequent decisions. The response phase is considered over when the threats to life, property, and environment are under control.

Advances in communication and information technology are providing emergency response teams with increasing amounts of data. Monitoring devices, sensors, and mobile communication provide the basis for operational support. It was previously mentioned that these technologies permit response personnel to "see" inside a damaged reactor with mini-video, "feel" the conditions inside a burning container with robot devices, and "track" the dispersion of pollution with satellite imagery. However, these advanced technologies only assist the response personnel in *sensing*; support is required for *reasoning* as well in the decision making process.

The operational risk management paradigm takes into account the uncertain nature of response activities – trucks may be unavailable, a sudden change in the weather may occur, or chemical dispersants may not function as expected. It also recognizes that this uncertainty may change the risk associated with various courses of action, for example, the fire may overrun the barricade. Although the approach supports the manager's decision making process, his or her role is not defined. Human cognitive limitations in operational environments must be considered as a constraint; consequently, our approach assumes that the human is always an integrated part of the decision making process.

1.2 Operational Risk Management in Emergency Management

The example addressed in this section is based upon the response to the Exxon Valdez oil spill in Alaska, USA [Harrald et al., 1990]. Responding to a maritime casualty where the release of the cargo could have potentially catastrophic impact calls for the establishment of a command and control center. The on-scene or incident commander and her or his staff gather data on the response activities and the impact of the incident. They also solicit expert judgment and knowledge, and monitor and control the implementation of the actions decided upon for the response operations. External events, particularly the weather, require that this monitoring and control be done in real-time.

Once an accident occurs, the incident command center is activated. It gathers data on such factors as type of cargo, type of vessel, extent of the accident, and weather, and decides on a course of action – in a very time-constrained situation. The validity of the chosen course of action depends upon the quality of the information available. In addition, the initial decisions act as constraints on subsequent decisions. If, for example, a decision is made not to use chemical dispersants or not to burn off the oil on the water, booms must be allocated to protect vulnerable resources. The response phase is considered complete when there is no threat to human life and the vessel and its cargo are under control.

The attributes chosen for assessing the preference of each activity are operational risks (R), operational costs (C), and value of the response activity (V), for example, the benefit of an activity that mitigates long term environmental damages. Highest priority is given to the safety of personnel on the vessel and in the emergency response team, and the population that may be affected by the release of the cargo. In the case of the Exxon Valdez, the humans at risk were those on the vessel. After the safety of personnel, the emergency response managers must consider environmental damage to the shoreline and the waters, property loss to the shipper, and income loss to fishermen and the tourist industry (now and in the future). The ability of each of the response activities to lesson these impacts is delineated in the class "Values."

Operational risk management for emergency response consists of three major components: the large scale operational system, the human-machine control system and the communication system (see Figure V.1). The large scale operational system is the environment where the actual response activities take place.

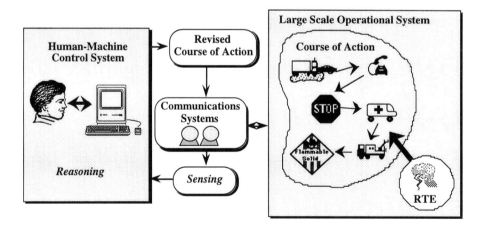

Figure V.1: ORM environment for emergency management.

The activities follow a planned CA. However, real-time events (RTEs), such as weather changes or the breakdown of equipment, can affect the anticipated safety and effectiveness of the planned CA. Advanced satellite-based communication systems monitor the performance of activities and the occurrence of RTEs. The human-machine control system does sensing and reasoning. Sensing is supported by the communication systems and reasoning by advanced computer technology. In this way, the human cognitive process needed for problem solving and decision making has technological assistance in the form of data bases and algorithms with both numeric and symbolic computational models.

The generic tasks of the control system consist of **sensing** and **reasoning** needed to manage multiple RTEs. RTEs occur unexpectedly and can last for a period of time of unpredictable length. For example, a fire truck might break down and be unavailable for a few hours. At the moment the fire truck breaks down, the planned CA must be revised if it requires this specific fire truck. As soon as the truck is repaired, the new course of action must be reexamined and revised as needed to reflect the availability of the vehicle. The state of the large scale operational system where no RTE is present and the planned CA can be executed is called **normal condition**; the state when an RTE is present is called **RTE condition**.

In real-world emergency response situations, several courses of action are monitored simultaneously. Whenever an RTE occurs, the ongoing courses of action must be assessed. Those that are affected by the RTE (i.e., those that are affected by the impact of the RTE) must be revised. When an RTE is resolved, the courses of action must be reassessed. Courses of action might also be initiated while RTEs are in effect. When these RTEs are resolved, reassessment must be done not only for the affected courses of action but also for all CAs that were initiated in response to the RTEs. Managing the response to the RTEs by formulating, deciding upon, and implementing CAs is a very complex and time consuming process. Only a perfect integration of the cognitive capabilities of the operational risk managers and the speed and reliability of communications and computing technologies can ensure effective emergency response.

The time pressure under which this control system must operate is due to the variable time frames for decision making. An RTE might require a particular course of action, which does not have to be initiated immediately. For example, after vessel grounding, the unavailability of equipment designed to control the dispersion of oil on the water might not be an issue if little oil has been released. The incident commander has some time to gather and process data on alternative means of containing and removing oil that might be released into the water. However, waiting to receive more information may eliminate possible courses of action, since alternate equipment and associated personnel will have to be activated and transported to the site. Therefore, the commander has to weigh the value of additional information against the need to make decisions in time to effect the desired response. The operational time to make decisions depends on what response activities have been initiated and are underway; this can only be determined by monitoring the courses of action and their impact in real-time.

2. ORM Decision Logic for Emergency Response

2.1 The Graph Model

Traditional approaches to emergency management are based on normative principles for selecting and implementing appropriate activities, such as for air traffic control [Schutte et al., 1987], vehicle navigation [Karimi and Krakiwsky, 1988], mission planning [Beaton et al., 1987], and process control [Moore et al., 1988], [Naum et al., 1989]. They do not include the integration of the operator to revise emergency plans to respond to external RTEs and the need to formulate and compute new optimal CAs. On the other hand, it must be remembered that operational performance decisions are more likely to be pre-structured than strategic planning decisions [Sage, 1986].

The prestructuring of operational decisions must be extended beyond the strategically devised contingency plans. The meta-reasoning structure must provide an environment where the operator can flexibly assess the impacts of RTEs on selected activities. In fact, the meta-reasoning structure proposed in this chapter allows one to assess the impacts of RTEs and reevaluate CAs at any time. This is accomplished by embedding the decision support paradigm into a visual interactive system that allows the operator to monitor the ongoing operations. In addition, and here lies the radical departure from traditional approaches, the commander, emergency response person, or operator can access any activity and change its value in regard to the successful completion of the emergency operation. Whenever the value of an activity has been changed, the embedded reasoning logic upgrades the CA.

Note that a CA consists of a temporal ordered sequence of decisions and concomitant selected activities. As shown in Figure V.2, each activity is represented as the link between two decision nodes. The first node is the decision to initiate the activity and the second node is the decision to initiate the next activity. The graph representation is necessary since most response activities cannot be initiated until other activities have taken place. For example, equipment may be available to contain oil on the water but the barge needed to transport the equipment may not be. The first node of the graph represents the decision to activate the emergency response plan. Once the response is underway, the first decision is to ascertain if it is necessary to reassess the plan due to RTEs.

The final node represents the decision to terminate the response to the emergency and initiate the recovery plan.

The decision/activity graph with oriented links is a **topological graph structure**, as defined in Chapter II. A feasible CA consists of a sequence of activities on the graph. This sequence begins at the first decision node, goes along some intermediate decision nodes (considering the orientation of the activities), and ends at the end node.

Reaching the destination node means that the response can be terminated and recovery initiated. However, not only are there several ways (courses of action) to reach a specific destination node, there also are several destination nodes calling for different approaches for recovery. Once the origin node has been identified, the search for an optimal CA becomes a single-origin/multiple-destination search problem on a graph.

Under normal conditions, there will always be at least one successful CA. However, under RTE conditions, there might not be any feasible CAs. The determination of all feasible walks (CAs) is a combinatorial problem on the topological graph, as discussed in Chapter II, Section 1.2. Depending on the number of activities and decisions, finding all feasible CAs can become a computationally intensive task.

2.2 Values of Activities and Computing Optimal CAs

The commander of the emergency response operations is not interested only in the **feasibility of activities** but also in their relative preferences. The preferences are described by a set of attributes that are relevant to the commander. The preference of two consecutive activities is then the "sum" of their preferences, and the preference of a CA is therefore the "sum" of the preferences of its activities. The best CA is the feasible CA with highest preference. Finding the most desirable (i.e., optimal) CA is done by an appropriate graph-theoretic algorithm, as discussed in Chapter II.

The most common attributes to be considered in emergency response are operational risks, values of activities, and costs. Operational risks refer to human life, environmental pollution and destruction, property loss (also expressed in costs), loss of company image, etc. The value of a response activity is, for example, the value of the mitigation of long term environmental damages. Operational costs are expenditures for an activity, which are assumed to be estimated with sufficient accuracy to be considered deterministic.

Some activities in emergency response are considered feasible but have risks so great that there is no monetary equivalent. On the other hand, there are also activities with low risks that do not justify the expenditures needed to abate or avoid them. Therefore, risks and costs have the following ordinal relation: High Risks $\prec$ Costs $\prec$ Low Risks, where "$\prec$" means "less desirable than."

The value of each activity refers to the ability of each of the response activities to lessen the emergency situation. If an activity is of any value, its costs are never too high to prevent its use in emergency response.

In addition, two border-classes are also introduced. Activities that do not have any significant risks, costs, or values will be assigned to the first class (ω); activities that under no circumstances are to be used will be assigned to the second class (α). Thus, the five classes have the following ordinal relation:

$$\alpha \prec \text{High Risks} \prec \text{Values} \prec \text{Costs} \prec \text{Low Risks} \prec \omega.$$

The feasible activities are therefore those that have a preference higher than α. The operator can assess all planned CAs that are depicted on the graph structure. Preferences of activities, just like their feasibility, can depend on previously implemented activities. For example, using chemical dispersants to respond to an oil spill has a lower risk (higher safety preference) if the effectiveness of the dispersants has previously been tested. After all activities have been assessed, the resulting graph is called a preference graph.

Whenever an RTE occurs, the operator must reassess the risks, values, and costs of all activities affected by the RTE. All the affected activities are presented to the operator one-by-one or grouped into similar operations for reassessment.

The risk, value, and cost elements that the preference of an activity (or a course of action) can assume are the following: Risks$\in \{\alpha, \text{high}, \text{low}, \omega\}$, Values$\in \{\text{indispensable, some}, \omega\}$, and Costs $\in \{\alpha, \$x, \omega\}$. An example of a preference of an activity is the following: [0,0,1,34,5,0]; which means that there are zero α, high risk, and ω elements, one high risk, 34 costs, and 5 low risk elements. Adding two preferences is done by adding the corresponding elements of the two preference vectors; for example, [0,2,1,3,6,1] $\oplus$ [0,3,1,3,1,0] = [0,5,2,6,7,1].

The optimal CA under normal conditions, as well as any changes in the CA due to an RTE are determined by an algorithm that operates according to the following hierarchical priorities:

(1) never take links with α Risks or Costs and take all links that have indispensable value;
(2) avoid links with high Risks as much as possible;
(3) take links with some Value;
(4) minimize Costs; and
(5) avoid links with low Risks as much as possible.

With these principles for determining optimal courses of action, it follows that $k\otimes\alpha=\alpha$ (non feasible). Furthermore, since ω is not used, it can be concluded that $k\otimes\omega=\omega$ (highest possible preference).

The relation between two decisions d_i and d_j (i.e., activity a_{ij}) is usually dependent on previously taken activities, decisions, and general conditions about the environment. Thus, the relation between two decisions is presented in the following way: ([Path],[d_i,d_j],Π_{ij},[History]). The Path is merely informational and is used to keep track of the CA. π_{ij} is the preference of the relation; it consists of risk, value, and cost preferences. The History is the condition required for the relation between d_i and d_j, the activity a_{ij}. It can consist of past decisions and activities as well as state conditions such as type of vessel, weather, type of cargo, etc. d_i is the predecessor decision and d_j is the successor decision. The connection of two relations to a CA is done by the algorithm in the following way.

Let ([],[A,B],Π_{AB},[F,G]) be the first relation and ([],[B,C],Π_{BC},[F,G]) be the second relation. Since both relations have the same history (F,G), they can be connected to the new relation ([A],[B,C],$\Pi_{ijAB}\oplus\Pi_{BC}$,[F,G]); that is, there is a CA from A to B to C with total preference of Π_{AB} "plus" Π_{BC}. If the relation ([],[C,D],Π_{CD},[F,G]) is added, then the following relation can be concluded: ([A,B],[C,D],$\Pi_{AB}\oplus\Pi_{BC}\oplus\Pi_{CD}$,[$F,G$]). To extend this relation, one has to look for a relation between D and any other decision that holds under F and G.

If A is the first decision and O the last decision of an emergency response operation, then a feasible CA has the following structure: ([$A,X_1,...,X_{n-1}$],[X_n,O],Π_{AO},[$Y_1,...,Y_m$]), where the X_i's stand for the intermediate decisions and the Y_j's for the history. The most desirable CA is the one with the highest preference Π_{AO}. A path from the origin decision to any other decision that is not the last decision is called an intermediate CA (ICA).

The principles to construct CAs are the following:

(1) connect an ICA with a relation if the relation's predecessor decision is identical with the ICA's successor decision and if (i) either they have the same history, or (ii) the relation's history is contained in the set Path∩History;

(2) if (1) holds, the new ICA's path is the old ICA's path plus the relation's predecessor decision, with the preference being equal to the sum of old ICA and relation preferences and the history being the old ICA's history∩relation's history.

2.3 Revising Courses of Actions

The preference of every activity must be assessed for expected operational conditions (normal situation). In addition, preferences must also be determined strategically for all those infrequently occurring events that are considered to be significant. This implies that an activity can have several preferences with different histories. In order to avoid ambiguity in the computation of optimal CAs, the preferences must be ranked according to their importance. Ambiguity arises when a history is a subset of another history. In this case, both preferences hold. Theoretically, such ambiguity could be avoided if for all activities, and for all theoretically possible histories (which are obtained by exhaustively combining decisions, activities, and state conditions), a preference is assessed. However, this would not be practically feasible. It is therefore suggested ranking a limited set of preferences for every activity.

After all activities have been assessed strategically, the optimal course of action can be computed for normal conditions. Since some deviations from normality also have to be considered strategically, the system has the capability to automatically adjust the CA to those conditions.

Usually, there are multiple end-nodes that lead to a satisfactory state where the emergency operation ends and recovery measures start. Thus, the search problem on the preference graph is to find all CAs from one origin to multiple destinations. The optimal CA is the one with highest overall preference.

As soon as an RTE occurs, the operator must try to match the RTE with a set of RTEs for which the impact could be strategically determined. If a match can be found, a new optimal CA is computed automatically from the last decision made to one of the destination nodes. Since CAs to different destination nodes might have different preferences, a minimum preference (level of achievement) for the operation must be defined. The feasible CAs are those that have a preference π_i that fulfills: $\pi_\alpha \gtrsim \pi_i$, where π_α is the level of achievement.

If no match can be found for an RTE, the preferences that are affected by this event must be reassessed by the operator. To do this, an algorithm must first identify the affected preferences. A preference is affected by an RTE if its predecessor decision, its successor decision, or an element of its history is affected. To determine the affected preferences, the operator must first determine which activities and state conditions are affected by the RTE. Then an algorithm determines all activities that are sensitive to the RTE. For those activities, the operator determines at least one new preference with the RTE being automatically part of the its history. This preference is then assigned the highest rank. Then, new optimal courses of action can be computed. As soon as the RTE is over, the preferences with that RTE in the history are removed automatically and new optimal courses of action are determined. They start with the last activity undertaken when the RTE was over and end at one of the destination nodes. In some cases, the new CA might be identical to the CA prior to the RTE. In this case, the strategically devised contingency plan could be resumed. An example is an equipment failure RTE that could be rapidly repaired.

Delays are special cases of RTEs. Although the average time needed for each is fairly well known, delays are commonplace in emergency operations. However, the impacts of most delays can be planned strategically and the system can upgrade the optimal CA automatically without interference by the operator. If time delays require a change of CA, the operator could be warned automatically.

2.4 An Illustrative Example

An example of the oil spill response activities is shown in Figure V.2. The dark lines represent the optimal CA before the RTE occurred. The gray lines show the change in the CA in response to the RTE.

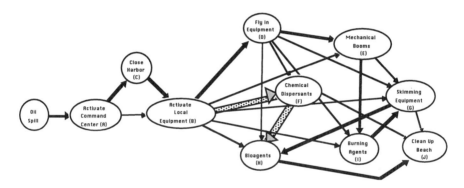

Figure V.2: Topological graph of oil spill response plan.

The preference structure was developed using the classes high-risk (h), values (v), costs (c), and low-risk (l), where values are measured by a surrogate for positive impact on the environment, like barrels of oil removed from the water or the shore. It is based upon assessment made in anticipation of the event, which is part of the oil spill response plan. These contingency plans are developed based on an expectation that once the event occurs, response activities can be carried out with certainty. It is our contention (and was the case in the Exxon Valdez accident) that this expectation is unrealistic, calling for operational risk analysis. The preference structure for the Exxon Valdez response is the following:

$$\alpha \prec h \prec v \prec c \prec 1 \prec \omega.$$

Both the assessment of the preferences and the choice process are illustrated with a numerical example. The following a priori preference relations have been defined, where the letters refer to the circles in Figure V.3, and $\pi(A\Re B)=2c$ means that the preference (π) of the relation ($\Re$) or link between decision A and decision B corresponds to 2c (2 units of cleanup costs). Preferences for the activities are assigned based upon our experience planning for and responding to major oil spills, such as the Exxon Valdez accident [Harrald et al., 1992].

$\{\pi(A\Re B)=2c; \pi(A\Re C)=10c,1; \pi(C\Re B)=c\}$

If A then:
$\{\pi(B\Re D)=c; \pi(B\Re E)=4c; \pi(B\Re F)=h,2v,5c; \pi(B\Re G)=v,c; \pi(B\Re H)=c; \pi(B\Re I)=c\}$
If A and C then:
$\{\pi(B\Re D)=c; \pi(B\Re E)=3c; \pi(B\Re F)=h,4v,2c; \pi(B\Re G)=v,c; \pi(B\Re H)=c; \pi(B\Re H)=c\}$

$\{p(D\Re E)=8c\}$

If A then: $\{\pi(D\Re F)=h,2v,10c; \pi(D\Re G)=v,5c; \pi(D\Re H)=2c; \pi(D\Re I)=2c\}$
If A and C then: $\{\pi(D\Re F)=h,8v,4c; \pi(D\Re G)=v,5c; \pi(D\Re H)=2c; \pi(D\Re I)=2c\}$

If A and D then: $\{\pi(E\Re G)=4v,5c; \pi(E\Re I)=h,v,2c\}$
If A and B then : $\{\pi(E\Re G)=2v,c; \pi(E\Re I)=h,3v,c\}$
If A and C and D then: $\{\pi(E\Re G)=12v,5c; \pi(E\Re I)=4v,c\}$
If A and C and B then: $\{\pi(E\Re G)=6v,c; \pi(E\Re I)=v,c\}$

If A and B and E then: $\{\pi(I\Re G)=h,2v,c\}$
If A and C and B and E then: $\{\pi(I\Re G)=4v,c\}$
If A and B then: $\{\pi(I\Re G)=h,v,c\}$
If A and C and B then: $\{\pi(I\Re G)=v,c\}$
If A and D and E then: $\{\pi(I\Re G)=h,2v,c\}$
If A and C and D and E then: $\{\pi(I\Re G)=8v,c\}$
If A and B and D then: $\{\pi(I\Re G)=h,v,c\}$
If A and C and D then: $\{\pi(I\Re G)=2v,c\}$

$\{\pi(G\Re H)=v\}$

If I then: $\{\pi(G\Re J)=v,10c,1\}$
If $\neg$I then: $\{\pi(G\Re J)=v,2c,1\}$

$\{\pi(F\Re J)=c,1\}$

$\{\pi(F\Re H)=1\}$
If D and G then: $\{\pi(H\Re J)=8v,c\}$
If D and $\neg$G then: $\{\pi(H\Re J)=4v,c\}$
If $\neg$D and G then: $\{\pi(H\Re J)=v,c\}$
If $\neg$D and $\neg$G then: $\{\pi(H\Re J)=v,c\}$

If F and $\neg$D then: $\{\pi(H\neg J)=v,c\}$

The best response for this particular spill situation, that is, the optimal path through the graph, is to close the harbor, fly in additional equipment, use booms, use burning agents (which requires the use of skimmers), and add bioagents, as shown in Figure V.2. The solution was determined by first enumerating the paths as shown in Figure V.3, and then "adding up" the preferences. The dark lines represent the optimal course of action before the real-time event. The gray lines show the change in the course of action in response to the real-time event.

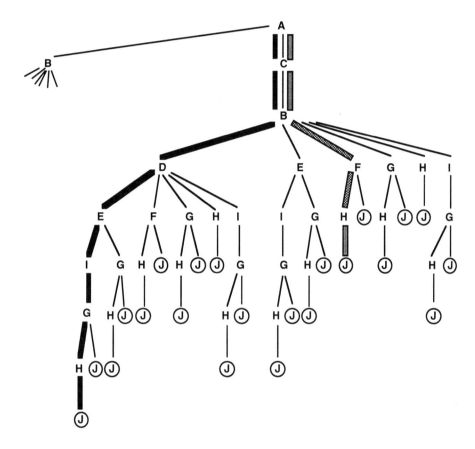

Figure V.3: Tree structure to enumerate paths in graph of oil spill response plan.

Note that the benefits in terms of gallons of oil removed are the same whether or not burning agents are used; however, burning agents are more cost effective. Since this analysis is done in the strategic phase of planning the oil spill response, it can be discussed whether closing the harbor was sufficient to remove burning from the high risk category; that is, if A and C and D (harbor closed) then: $\{\pi(E\Re I)=4v,c\}$ versus If A and D (harbor open) then: $\{\pi(E\Re I)=h,v,2c\}$.

Now let us assume that weather prevented the response equipment from being flown in and the skimming equipment on site was not operating; that is, a real-time event. The links to be reassessed are as follows:

If A then: $\{\pi(B\Re F)=2v,10c;\pi(B\Re H)=c\}$
If A and C then: $\{\pi(B\Re F)=8v,4c;\pi(B\Re H)=c\}$

$\{\pi(F\Re H)=1\}$

$\{\pi(F\Re J)=10c,1\}$

If F then: $\{\pi(H\Re J)=4v,2c\}$
If ¬F then: $\{\pi(H\Re J)=v,c\}$

The revised course of action, the new optimal path, is to close the harbor and use both chemical dispersants and bioagents, that is, "use everything we've got", as shown in Figures V.2 and V.3. The reason for this result is that environmental regulations prevented the use of burning agents without skimming, that is, that link was assessed to fall into the class α a priori and is, therefore, not feasible.

3. Decision Support System

3.1 The Modeling Environment

The graph-theoretic paradigm of operational risk management has been implemented into a prototype decision support system in a multimedia environment. The design of this prototype system is based on the discussions in Chapter III.

This prototype decision support system can operate at two different levels. At the first level, it works as a development shell for modeling the system. Within this developing environment, the system supports the construction of the relational structure of the emergency operation. In a visual interactive mode, nodes and arcs can be defined, deleted, and rearranged by choosing the corresponding mode (Figure V.4). A predefined set of icons helps visualize the decision nodes. A decision node is defined by clicking on the decision box and then choosing one of the predefined icons. An oriented arc between two decisions is defined by clicking on the arc box and then clicking on the origin node followed by the destination node. The removal of a node or arc is done first by choosing the modes on the right hand side of the menu box and then repeating the procedure that was necessary to create them.

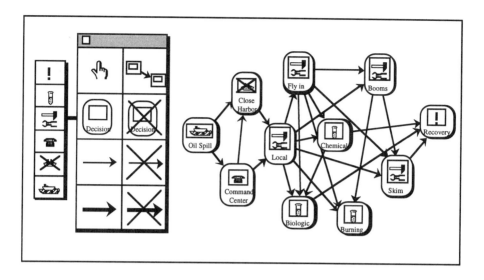

Figure V.4: Editor mode for graph model.

For every activity (arc between two decision nodes), a prestructured window for assessing the preferences is reserved. The operator can assess the preferences, the state and activity histories and the sensitivities. When multiple preferences must be defined for one activity (which is usually the case), the operator is asked to rank them. The reassessment of activities, after an RTE has occurred, is done using the same window.

The second level at which this prototype operational decision support system can be used is the run level. With the assessed activities, an optimal course of action can be computed and the solution highlighted on the graph model. As soon as the operation starts, the current activity is highlighted. A timer shows the expected remaining time for this activity. Any deviations from what is really going on in the operation are recorded. If time delays cannot be handled by the system automatically, the operator is called to fill the gap.

3.3 Preference Assessment and Decision Making

The prototype system allows the simulation RTEs, both single and multiple. Whenever an RTE occurs, the operator must assess the effects of the RTEs so that the system can determine the affected activities and preferences. The system does this by checking which activity is sensitive to the impact of an RTE. Then, the affected activities are presented to the operator one-by-one for reassessment as shown in Figure V.5. The new preferences are added to the old preferences with highest priority.

After the assessment of the preferences is done by the operational risk manager, a new optimal CA is computed. Both the new as well as the strategically planned CA are highlighted on the graph model. The operational risk manager can then choose among keeping the planned contingency plan, taking the new one computed by the system (based on his/her assessment), or even taking a different one by selecting new activities with mouse clicks on the screen.

As son as the RTE is over, the system checks whether or not the operation should return to the originally devised contingency plan.

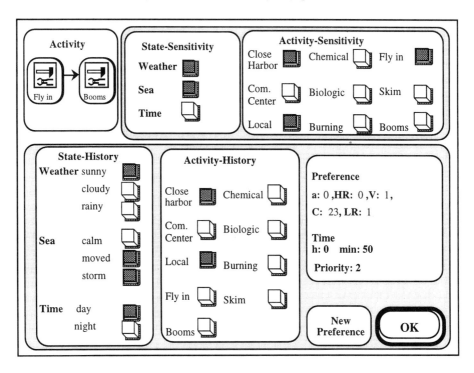

Figure V.5: Assessment window for activities.

The example currently implemented in the prototype system is based upon the response to the Exxon Valdez oil spill in Alaska, USA [Harrald et al., 1990]. As noted, the commander at the incident response center and the staff gather data, do analyses, make decisions and monitor their implementation. These activities must be done in real-time since RTEs (e.g., weather changes) can occur at any time, having a major impact on the planned activities.

At the beginning of the response operation, the incident response center gathers data on type of cargo, type of vessel, extent of the accident, and weather. With these conditions and the fact that an emergency situation has occurred as history, the initial course of action is computed.

The attributes and the ordinal relation is the same as previously discussed: $\alpha \prec$ High Risks $\prec$ Values $\prec$ Costs $\prec$ Low Risks $\prec \omega$. In the case of the Exxon Valdez, the humans at risk were those on the vessel. The highest priority is given to their safety. At the second level, environmental damage to the shoreline and the waters, property loss to the shipper, and income loss to fishermen and the tourist industry (now and in the future) are considered.

All assessed preferences can be accessed and changed. RTEs can be simulated and terminated at any time. The system was developed in Hypertalk on a Macintosh. Algorithmic procedures were programmed in Pascal and included as external commands. This assures that the system is fast and efficient enough to be used by experts in the field.

3.4 Conclusions

The decision logic for operational risk management was extended to operational situations where there is no natural graph structure like a road network. The two applications discussed in the next Chapter are air raid command and nuclear power generating management. They demonstrate the generalizability and the feasibility of operational risk management.

CHAPTER VI: OPERATIONAL CONTROL

This chapter demonstrates the generalizability of operational risk management (ORM) by presenting two additional applications: command and control of air force raid and emergency response to an incident at a nuclear power generation facility. In both cases prototype decision support systems (DSS) were built that employ the ORM logic and hypermedia technology. The air raid command DSS was designed to support the individual commander, while the nuclear power emergency response extends the ORM logic to support multiple groups of managers.

1. Air Raid Command
1.1 Introduction

This section discusses RARC, a prototype decision support system for air raid command and control. RARC supports the decision making process of an en-route commander in a fighter-bomber who is in charge of 10 to 20 airplanes. In this hostile environment, strategically planned courses of action must be changed on a continuous basis due to unexpected activities by the enemy.

The RARC system can suggest different tactics when an enemy interceptor or surface-to-air missiles are detected. Because these tactics affect the flight plan, estimated fuel consumption and estimated time of arrival must be revised. Data from airborne warning and control systems are used for situation assessment during air-raid operations.

The structure of RARC is based on the paradigm of operational risk management, as proposed in Chapter II and applied in Chapter IV for transportation of hazardous materials and Chapter V for emergency response.

1.2 The ORM Environment

In this application, the ORM environment consists of three components:

1) the large-scale operational system (LSOP), which includes the air raid fighters and bombers; LSOP operates according to strategically designed courses of action (CAs);

2) the real-time events (RTEs), that is, the emergency incidents, which have an impact on the expected achievement of the planned courses of action, and which cannot be considered strategically; and

3) the human-machine system that (i) monitors the LSOP as part of passive ORM, (ii) assesses impacts of RTEs on the planned CA as part of active ORM, and (iii) if necessary revises the CA as part of active ORM.

Air raiding, as conducted by the Japanese Air Defense Forces, consists of an air-raid group assigned to bomb a target in enemy occupied area. The group has about ten fighter-bombers and ten escorting fighters, with the commander in a fighter-bomber. The target area is covered by an area-control radar aircraft, such as an Airborne Warning And Control System (AWACS). The AWACS is a central information controller which gathers data in real-time about the operations and the environs and provides reports to the commander. The data are obtained from the AWACS' on-board radars which give the commander information about enemy activities, such as the position of enemy interceptors and enemy Surface-to-Air Missiles (SAM); SAMs try to interfere with the air raid mission.

Prior to takeoff, the commander is given information on the potential positions of SAMs and the route of enemy interceptors. With this information, the commander determines the flight plan. Although the flight plan avoids threat areas, the enemy might be encountered in unexpected situations (RTEs).

The commander makes decisions in real-time to reduce risks. The commander's task is to direct the group to the assigned target safely and with minimum costs (e.g., fuel consumption). More specifically, the commander must: (1) assign escorting fighters to cope with enemy interceptors, and (2) select alternative flight plans in order to avoid enemy SAMs.

The only way to reduce the threat from enemy interceptors is to deploy friendly fighters to destroy them. The AWACS can detect an enemy interceptor's position, speed, heading, and size. The size is given as "Single," "Few," or "Mass," where one interceptor is regarded as "Single," two or three as "Few," and "Mass" is more than four. Data about real-time events (RTEs), that is, enemy interceptors, are sent to the commander.

SAMs can be classified as "Infrared" or "Radar" guided missiles. AWACS can detect radar-guided missiles when the missile's radar is activated. Infrared missiles cannot be detected by AWACS, but their range is very limited. Since missile launching sites can be moved, their exact positions cannot be obtained before the operation commences. While a missile's radar is silent, it is almost impossible to locate the missile's position. Intelligence provides data on the SAMs, including their radar frequency, range, effects, etc. Once a missile's radar starts to emit radar waves, AWACS can determine the type of SAMs using their radar emissions. AWACS notifies the commander of the emerging SAM (RTE), conveying its type and position. Effective ways to reduce threats from SAMs are: (1) changing the course of the group; and (2) climbing to a higher altitude in order to fly above the effective altitude of the SAMs. Another way to reduce the threats from SAMs is electric counter measure (ECM), where aircraft emit radio waves that jam enemy radars. ECM reduces the SAM radar's detecting range and is assumed to be activated when the group flies over enemy areas.

The commander is responsible for the success of the raid. Currently, AWACS can recommend a new flight path to avoid enemy activities; however, the decision must be made by the commander. Plans are underway to provide the commander with the same information as the controller in the AWACS. Unless support is provided to the commander, it may be difficult for that person to control his or her aircraft and, simultaneously, make decisions on the flight plan for the group. The purpose of this project was to employ operational risk management embedded into a decision support system with hypermedia to support the commander in this decision situation.

Due to advanced communications technology, the commander does not have to be located within the operational system. However, it is assumed that the loop between sensing and reasoning will always include humans.

1.3 ORM Decision Logic in Air Raid Command

As in the previous applications, courses of action (CA) can be represented on a graph structure, consisting of nodes (vertices) and links (edges). Nodes represent decisions and links are activities. As discussed in previous chapters, oriented link from decision node d_i to decision node d_j indicates that decision d_i has been made to undertake activity a_{ij} which leads to decision d_j; that is, the link a_{ij} states that there is a relation between the two decisions d_i and d_j. Therefore, every activity leads to a subsequent decision, except for the last decision, that is, that normal operation can be resumed.

The commander is interested not only in the feasibility of activities but also in their benefit, effectiveness, or preference. Preference is described by a set of attributes that are relevant to the commander. The preference of two consecutive activities is equal to the "sum" of the two preferences, and, therefore, the preference of a CA is equal the "sum" of the preferences of its activities. The best CA among the feasible CAs is the one with highest overall preference. To find the optimal or most desirable CA, a graph-theoretic algorithm is used as discussed in previous chapters.

The most common attributes to be considered in an air raid assault are operational risks and costs. Operational risks refer to human life, loss of material, etc. Operational costs are expenditures for an activity; that is, expected fuel consumption to the target and estimated time of arrival. Costs are assumed to be estimated with sufficient accuracy that they may be considered deterministic.

Some activities in an air raid may be feasible but with risks that exceed any monetary value. Furthermore, some activities with low risks do not justify the expenditures needed to abate or avoid them. Therefore, risks and costs in an air raid have the following ordinal relation:

$$\alpha \prec \text{High Risks} \prec \text{Costs} \prec \text{Low Risks} \prec \omega.$$

where "$\prec$" means "less preferred."

For an activity to be feasible, its preference value must be greater than α. The commander can assess all planned CAs with this preference structure. For each RTE, the commander must reassess risks and costs of the activities which are affected by the RTE. For this purpose, all these activities will be presented to the commander one-by-one.

The risk and cost elements that the preference of an activity (or a course of action) can assume are the following: Risks$\in \{\alpha, x \text{ high}, y \text{ low}, \omega\}$, and Costs$\in \{\alpha, \$z, \omega\}$. An example of a preference of an activity is: [0,5,12,8,0], which means that there are zero α, 5 High Risk units, 12 Cost units, 8 Low Risk units, and zero ω preference units. Adding two preferences is done by adding the corresponding elements of the two preference vectors; for example:

$$[0,5,12,8,0] \oplus [0,2,45,6,0] = [0,7,57,13,0].$$

The optimal CA under normal conditions, as well as any changes in the CA due to an RTE are determined by the same algorithm discussed in previous chapters, and operates according to the following principles:

(1) it never takes an activity with Risk or Costs of value α;
(2) it avoids as much as possible links with High Risk values;
(3) it minimizes Costs; and
(4) it avoid as much as possible links with low Risks.

1.4 Decision Support System

A) The Concept

A prototype decision support system, RARC, was developed to aid the commander in an air raid. The system provides highly interactive interfaces, including a colored map with visualization and pseudo 3-D capability, in order to give the commander an image of the operational region and the en-route flight situation. RARC simulates all the features of the bombing operation. The flight plan and enemy activities, such as incoming enemy interceptors and emerging SAMs, can be displayed on the colored map. During the flight, the display uses pseudo 3-D graphics to show the current position of the group, including its altitude. The effective range of a SAM is shown by a circle around a SAM's symbol; and air raid group within the circle is in the SAM's effective range.

Figure VI.1 shows an operations window of RARC. The window displays not only the map, but also navigational information, such as altitude, air speed, heading, current fuel, and target information.

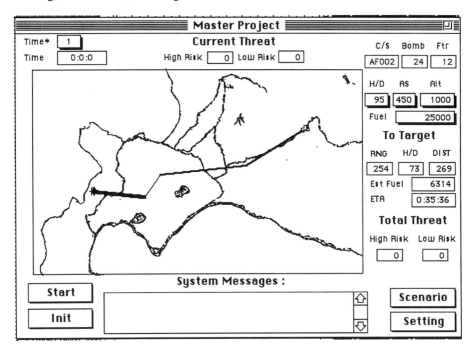

Figure VI.1: An operations window of the RARC.

When an RTE is detected, RARC provides the commander with several tactics along with their associated risks and costs (expected fuel consumption to the target and estimated time of arrival (ETA)). Using these assessments, the commander can choose a feasible or even the optimal tactic. RARC also warns the commander when the group reaches transition points, such as turning and climbing/descending, and when an RTE is detected. Thus, the commander can make decisions even while controlling an aircraft.

When incoming enemy interceptors are detected, their threats are assessed as High Risk, since they are very dangerous for the group. After friendly fighters are sent to intercept them, the risks are revised to Low Risk. The values of the risks of enemy interceptors are based on their size and number of friendly fighters deployed against them.

Table VI.1 shows the risk values when enemy interceptors are detected. RARC provides the commander with the tactics and their risk value in order to determine the number of the friendly fighters needed to cope with the incoming interceptors. If the group does not have any fighters, the system automatically assigns zero fighters, and shows its risks on the window. In this case, the commander may abort the mission, because the risks are too high.

# of fighters	0		more than 8		5-8		1-4	
Risk	High	Low	High	Low	High	Low	High	Low
Single	2	0	0	1	0	2	0	3
Few	3	0	0	2	0	3	0	4
Mass	4	0	0	3	0	4	0	5

Table VI.1: Risks of enemy interceptors.

Threats from enemy SAMs are regarded as Low Risks, because SAMs cannot destroy all the aircraft, and their effectiveness is much lower than that of enemy interceptors. Although Low Risk values of SAMs and their features should in practice be determined by intelligence, it was assumed that there are two types of SAMs. SAM1 is relatively old, with a range of 30 nautical miles and a single warhead. SAM2 is the latest missile system, which has a 50-mile range with a warhead big enough to destroy several aircraft at the same time. The possibility of a nuclear warhead on a SAM was not considered. Assumed values for Low Risk from the enemy SAMs are given in Table VI.2.

	No Action	High Altitude	Avoid the Area
SAM1	2	1	0
SAM2	4	2	0

Table VI.2: Risk values from enemy SAMs.

As previously noted, effective tactics to reduce the threats from SAMs include to climbing to higher altitude or avoiding effective areas of the SAMs. Table VI.2 shows the risk values associated with the tactics under ECM conditions.

Air raid tactics recommend that the group fly at high altitude while it is in friendly areas because of better fuel consumption, and drop to low altitude while in an enemy area enroute to the target. After the bombing has been completed, the group is expected to fly at a high altitude and to escape at high speed. This tactic is called "H-L-H." Flying at low altitude is safer, since it is harder for enemy radar to detect the group. A detected assault group may invite enemy interceptors.

B) Contingency Tasks

Once enemy interceptors are reported by AWACS, the commander should send friendly fighters of the group to destroy them. Assigning High or Low Risk value depends on the size of the incoming interceptors and the available tactics, that is, the number of friendly fighters sent to the enemy interceptors. The risk values are given in Table VI.1. After the commander decides on the number of fighters to send and enters it into RARC, the system displays the risk values on the operations window. The tactical options given to the commander are to send: (1) more than 8 fighters, (2) 5-8 fighters, (3) 1-4 fighters, or (4) no fighters. Each tactic is given with the risk values. Figure VI.2 shows the suggested tactics (number of fighters to be deployed) with the corresponding risk value.

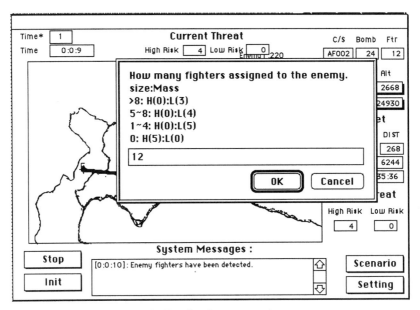

Figure VI.2: Tactics for enemy interceptors

RARC suggests four tactical options when an enemy SAM is detected. These are: (1) to avoid the SAM's effective area, (2) to climb to high altitude above it and descend after leaving the area, (3) to climb to high altitude and go straight to the target maintaining high altitude, and (4) to take no action. Since the tactics (1)-(3) change the flight plan, estimated fuel consumption and ETA to the target should be revised. Thus, the RARC shows the tactical options with their risks, the estimated fuel needed, and the ETA.

If tactic (3) is chosen, and if SAMs have already been detected, then the group may fly over the SAMs, and the risk should be increased accordingly. Even in this situation, the system calculates the risks of tactic (3) correctly. Since tactics (2) and (3) suggest flying at high altitude, the system automatically adds a Low Risk of 1 because of the "potential threat of high altitude flying." If the group is already flying at high altitude when a SAM is detected, tactics (2) and (3) are not applicable. RARC gives the commander only tactics (1) and (4). When the four tactics are displayed, they are presented in ascending order according to their risk values. Figure VI.3 shows four suggested tactics when a SAM is detected.

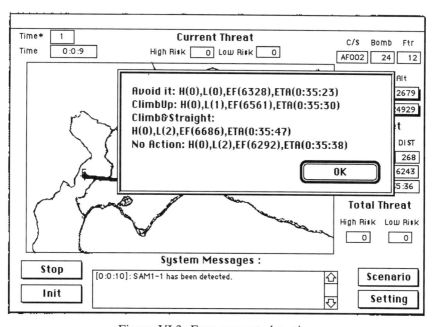

Figure VI.3: Four suggested tactics.

After the commander decides which tactic should be taken and enters the decision into the system, RARC shows a new flight plan on the map.

C) Display

RARC displays tactical situations and navigational information about the group. "Current Threat" shows the High and Low Risk values that the group is facing. "Total Threat" indicates the risk values for the entire operation at a given point in time. In "To Target," the system shows a straight range, a straight heading, the distance following the current flight plan, the estimated fuel, and the ETA to the target.

RARC uses three types of lines to show the group's flight plan or path. A thin line indicates the flight path with a descending leg, a medium line is for a level flight, and a thick line indicates a flight path with a climbing leg. The symbol "✦" indicates the ground position of the group, and the symbol "⌐" the 3-D position of the group. The system has pseudo 3-D capability in order to give the commander an intuitive image of the air space. The symbol "⌐⌐" shows the position of the target. Figure VI.4 shows the pseudo 3-D graphical display.

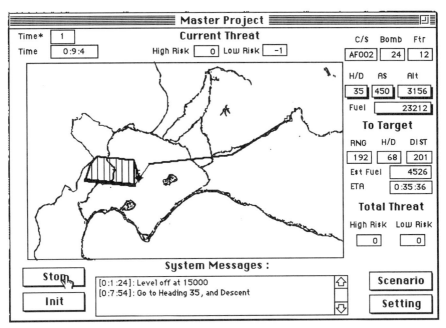

Figure VI.4: Pseudo 3-D graphical map.

The symbol "1" stands for a SAM. On the upper right of the symbol, information about a SAM is displayed including its type and its distance from the group. There is a circle around the SAM icon displaying its effective range. The symbol "E" indicates the position of enemy interceptors. The upper right portion of the symbol displays information about each interceptor's call sign, distance to the group, size, air speed, and heading on the upper right of the symbol. RARC uses radar data from AWACS, which is simulated by this system, to move the symbol as the interceptors advance.

D) An Illustrative Example

RARC obtains data from the AWACS via a data-link system. These data update the situation of the air-raiding operation to inform the commander of the tactical operation. In this example, the data are simulated. Figure VI.5 is a Scenario Setting Window. In this window, the original flight plan and the simulated data of enemy interceptors and SAMs can be entered into the system.

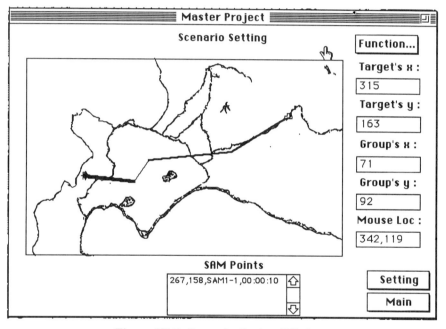

Figure VI.5: Scenario Setting Window.

The original flight plan should be determined using the information about provisional positions of enemy fighters and SAMs from the intelligence department. The flight plan is the optimal path (course of action) developed before the operation commences.

To simulate operations, RARC needs initial data for the group, such as its air speed, number of bombers and escorting fighters, etc. Figure VI.6 is the window used to enter the initial data on the group. With these data, the system simulates all possible situations faced during the operations. This input should be done prior to the air raid.

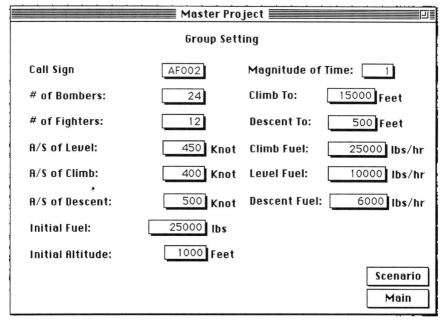

Figure VI.6: Group Setting Window.

E) Conclusions

Service providers of communications and information systems have developed generic monitoring and control systems. However, if one wants to tailor these systems to a specific task (such as air raid command), two major shortcomings become obvious. First, the commercial systems provide few analytical tools for decision support. Second, the systems are built with a closed architecture, which are difficult to alter. Consequently, if one wants to design and test novel human-machine control systems, different approaches must be investigated.

The RARC system is developed in a multimedia environment; it capitalizes upon latest achievements by using a hypermedia architecture. Activities are displayed during flight on a colored map with pseudo 3-D graphics. Virtual instruments are used as interfaces between the RARC system and the human operator. Animation is used to stimulate the commander's intuition and to support his/her decision making process.

The structure of RARC is based on the paradigm for operational risk management as discussed in Chapter II and operationalized for hazardous material transportation (Chapter IV) and emergency management (Chapter V). Operational risk management is formalized as a graph structure with nodes as decisions and the connecting arcs as the activities. Preferences for activities are assessed using a lexicographic preference order with non commensurable preference classes. The attributes used for RARC are High Risk values, Low Risk values, and Costs as determined by the incoming interceptors and the commander's tactics.

The prototype decision support system has been evaluated by commanders in an ad hoc manner, that is, sitting down and going through the system. Their comments included the need for a better input device, perhaps voice, and an increased resolution display. The next step is to conduct experimentation in an controlled setting (such as proposed in Chapter VII), with training exercises, conducted in simulators prior to implementation of the system.

2. Emergency Response at a Nuclear Power Generation Facility

2.1 Introduction

A major concern for an industrial operation is to know how to handle sudden onset incidents which occur very infrequently but which have potential detrimental consequences. The most important aspect of emergency management refers to the timely collection, processing, and dissemination of data, as basis for decision making. Traditionally, a set of alternative response plans are devised in a strategic manner. However, RTEs can occur for which no plan has been devised. In such instances, the operators must investigate, and possibly change, these strategically devised plans to prevent or mitigate the potential consequences. To optimize the operation and the response to RTEs, a functional task allocation between human and system must be devised. Table VI.3 shows such a functional task allocation as proposed by Hinz et al. [1993].

In large industrial organizations, however, the assessment and decision making process must be done rapidly under stressful conditions, involving different groups of experts. To cope with this combinatorial complexity of assessing and revising courses of action, the graph theoretical approach for real-time operational risk management as discussed in previous chapters will be revised to include multiple experts.

Process Control Tasks

- Control of energy transformation process
    start-up of systems, adjustment of setpoints, failure management
- Observing safety goals:
    confinement of activity, core cooling, ...

Task Category	
Human	**System**

- heuristic inference
  (impossibility of automation)
- flexibility
  (automation not economical)
	- large amount of data
	- monotonous tasks
	- fast execution requited
	- severe consequences/correction impossible in case of malfunction

Allocation of Tasks to	
Human	**System**

Process Control
- start/stop, optimization of processes
- setting of desired values
- monitoring/supervision of processes
- diagnosis of RTEs (preference assessment)
- manual intervention during RTEs
	Process Control
	- remote operation
	- short term/highly reliable automatic actions
	  • systems, component protection
	  • feedback control
	  • limitation system
	  • reactor protection system

Information Management - selection of graphics - acknowledgment of alarms	**Information Management** - to control tasks - to human capabilities for information perception/processing

Table VI.3: Functional task allocation [Hinz et al., 1993].

The focus of the next section is on providing effective decision support for groups of experts engaging in emergency response for large-scale industrial systems, with special attention given to the dissemination of information among the groups. Thereby, the assessments and choices made by the different groups cannot simply be aggregated to an overall organizational policy. For example, emergency response to an incident at a nuclear power generating facility involves at present three groups which are located at different sites, where communication between these groups is verbal with limited use of video [Hatcher, 1995].

2.2 The ORM Environment

Each group in an organization for emergency response is responsible for a unique set of tasks. In general, one group cannot overrule another group's recommendation or perform its tasks. The decision structures of emergency response organizations might have a single ultimate decision maker, who has the final responsibility for the actions taken in response to an emergency. In the process of communication and information exchange, however, the groups may generate several possible courses of action. Then, the groups decide together on how to rank the alternative courses of action for presentation to the decision maker as recommendations.

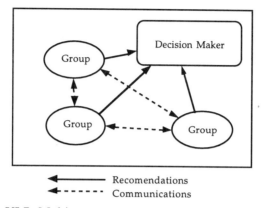

Figure VI.7: Multi-group emergency response organization.

A typical emergency response organization for large-scale operations can be divided into four groups consisting of one or more experts. The definition of the groups is based on the members' responsibilities in handling an accident as follows:

(1) Operators, who execute procedures of the operation under the decision maker's supervision in order to accomplish the objectives of the emergency response.
(2) A recommendation-generating group, which generates recommendations for the decision maker to consider. The members of this group have special knowledge of the technologies of the operations and generate recommendations based on their expertise.
(3) A management group, whose responsibility is to consider the impact on the organization responsible for the operations.
(4) A decision maker, who has the authority and responsibility to make decisions about the operations in an emergency situation. The decision maker will change with the increasing severity of the event. For purpose of this discussion, it will be either the operators or the management group.

The objective of the emergency organization and of the groups and their members is to resolve the emergency situation and to return the operation to a normal state. Therefore, any decision support technology must aid all the groups in: (1) monitoring, (2) real-time information exchanging, (3) interconnectedness, (4) real-time recommendation generation with preference values, (5) user-friendly interfaces, and (6) targeted information display.

To avoid information overload, the system displays must be designed to provide only the information appropriate for a particular group based on its responsibilities. In addition, the technology must include a modeling capability that provides the groups with the cognitive support needed to facilitate decision making in the stressful situation of emergency response.

2.3 ORM Decision Logic for Emergency Response at Nuclear Power Plant

A) Preference Structure

Real-time events have the potential to initiate an emergency and, in the course of an emergency, to make the emergency more severe and harder to control. It is therefore of major importance that the groups and their members be aware of all real-time events.

To provide effective decision support, a reasoning model must be devised which generates a set of recommendations in real-time and orders them according to the groups' preferences. The concept of the reasoning logic is described in Chapter II. A course of action (e.g., response plan) consists of a concatenated set of decisions and subsequent actions. A compact way to represent the courses of action and their relation is to use a graph structure, where the actions are the edges and the decisions to take the actions are the vertices. A specific course of action can therefore be seen as a "path" on the decision-action graph.

The preference for taking an action under normal conditions (i.e., as planned strategically) is assessed in terms of different attributes, such as risks to human, operational costs, and risks to the environment. These attributes should not be interpreted as evaluation criteria. Rather, they should be seen as preference classes which are related to each other by a complete strong preference structure, stating, for example, that high risks are of higher concern than operational costs. When a real-time event occurs, the preferences of the actions which are affected by the event must be reassessed. With these reassessed preferences, new optimal courses of action can be computed.

The reasoning logic for this application includes the following three components: (1) the large-scale operational (LSOP) system, which is supposed to behave according to strategically designed courses of action (CAs); (2) unexpected real-time events, which have an impact on the expected achievement of the planned courses of action and which cannot be considered strategically;

and (3) the human-machine system, which (i) monitors the LSOP as part of passive ORM, (ii) assesses impacts of RTEs on the planned CA as part of active ORM, and (iii) if necessary, revises the CA as part of active ORM. In addition, it adds a recommendation generating capability that ranks recommendations according to the emergency situation, including the potential for catastrophic impacts and the responsibilities of the users.

When the preferences for all activities have been assessed, the overall preference for a course of action, or parts of it, can be computed. The preference for two consecutive activities is equal to the "sum" of their individual preferences, where the sum is based on a lexicographic principle. That is, the preferences within a preference class are added up, but the preference classes remain incommensurable. This means, for example, that if high risks are of higher concern than operational costs, then there is no level of operational costs that could override concerns for high risks.

The preference of an entire CA is therefore the "sum" of the preferences of all of its activities. The best CA among the feasible CAs is the one with highest overall preference. Finding the most desirable CA is done by an appropriate graph-theoretic algorithm. The same five preference classes are used as in Section 1 for the air raid command case:

$$\alpha \prec \text{High Risks} \prec \text{Costs} \prec \text{Low Risks} \prec \omega,$$

where "$\prec$" refers to "less desirable."

For an activity or a course of action to be feasible its overall preference must be higher than α. When the preferences of all activities are assessed, the graph is called a preference graph. Optimal courses of action are determined by an algorithm that operates according to the following hierarchical priorities:

(1) avoid Risks or Costs of value α;
(2) minimize the number of High Risk values;
(3) minimize Costs; and
(4) minimize Low Risk values.

B) Preference Ranking with Multiple Groups of Experts

When multiple groups are involved in a distributed environment, each group might have a different preference structure to express preferences for actions, given the different responsibilities of the groups. Let's assume the following meta-preference structure consisting of the seven preference classes α, High Risk (HR), Low Feasibility (LF), Cost, Low Risk (LR), High Feasibility (HF), and ω:

$$\alpha \prec HR \prec LF \prec Cost \prec LR \prec HF \prec \omega.$$

Then, different groups could have different preference structures to use, such as given in Table VI.4.

	α	HR	LF	Cost	LR	HF	ω
Group1	*	*			*		*
Group2	*		*			*	*
Group3	*			*			*

Table VI.4. Responsibility for assessing preference classes.

The preference spectrum of Group1 is restricted to the preference classes $\alpha \prec HR \prec LR \prec \omega$. Thus, for Group1, the selection of the preference classes and their order is:

$\alpha \prec HR \prec - \prec - \prec LR \prec - \prec \omega$. which can be written as $\alpha \prec HR \prec LR \prec \omega$.

For the personnel in Group2, the preference structure is:

$\alpha \prec - \prec LF \prec - \prec - \prec HF < \omega$, which can be written as $\alpha \prec LF \prec HF$ v ω.

For Group 3 we have:

$\alpha \prec Cost \prec \omega$.

In unexpected deviations from normal conditions, the group members must assess the impact of the RTE on all affected activities using their designated preference structure. It can be assumed that the affected activities are presented to the experts one by one.

To illustrate the aggregation of the preferences across the groups, an example is given (see Table VI.5). Let's assume that Group1 recommends two alternatives with the following preferences: Alt. 1 = [0, 2, -, -, 0, -,0] and Alt. 2 = [0, 2, -, -, 0, -, 0]. Let's assume that Group2 assesses the preferences of the same two alternatives as follows: Alt. 1 = [0, -, 1, -,-,3, 0] and Alt. 2 = [0, -, 1, -,-,4, 0]. Let's assume that Group 3 assesses the alternatives as: Alt. 1 = [0, -, -, 5,-,-, 0] and Alt. 2 = [0, -, -, 4,-, -, 0], where the preference functions are monotonically decreasing, such that a $5 cost alternative is less desirable than a $4 alternative.

For Group1, the two alternatives have the same preference order, while Group2 prefers Alt. 1. On the other hand, Group3 prefers Alt. 2.

To aggregate the preferences across the groups, the meta-preference structure is used which defines the preference relations of all the preference classes used by the different groups: $\alpha \prec HR \prec LF \prec Cost \prec LR \prec HF \prec \omega$.

		α	HR	LF	Cost	LR	HF	ω
Group1	Alt. 1	0	2	-	-	0	-	0
	Alt. 2	0	2	-	-	0	-	0
Group2	Alt. 1	0	-	1	-	-	3	0
	Alt. 2	0	-	1	-	-	4	0
Group3	Alt. 1	0	-	-	5	-	-	0
	Alt. 2	0	-	-	4	-	-	0
aggregated assessments	Alt. 1	0	2	1	5	0	3	0
	Alt. 2	0	2	1	4	0	4	0

Table VI.5. Assessments of the alternatives by each group.

The aggregated assessments define the preference order of the two alternatives to be: Alt. 1 $\prec$ Alt. 2. The decision maker now has vital information on all the groups' preferences for the alternatives, as well as on the aggregated preferences. However, to implement this proposed decision logic, the groups involved in the decision making process must be supported in their communication process in real-time. As discussed in previous chapters, this should be accomplished by the use of hyper- and multimedia technology.

2.3 Decision Support System

A prototype decision support system (DSS) has been developed that incorporates the proposed decision logic and multimedia technology for an emergency response organization in a nuclear power generation plant. The prototype DSS supports three sites: the Control Room (CR), the Technical Support Center (TSC), and the Emergency Operation Facility (EOF). The connections among these sites allow all the members of the Emergency Response Organization (ERO) to share vital information. The operators and the operational decision maker on duty are located in CR. The recommendation-generating group is housed in TSC, while the management group stays in EOF.

The connections among these three sites have been established using the QuickTime Conference technology, including video conferencing systems, the capability to exchange command messages, and several other information exchange capabilities. All the multimedia data is contained in one individual workstation and controlled with the command messages via QuickTime Conference linkages. This scheme reduces the amount of computer traffic on the computer network.

A) Functions of the Prototype System

Figure VI.8 shows an actual view of the prototype system supporting the Control Room during normal or routine operations, that is, when an emergency has not been declared. Thus, communications links to the other sites are dormant.

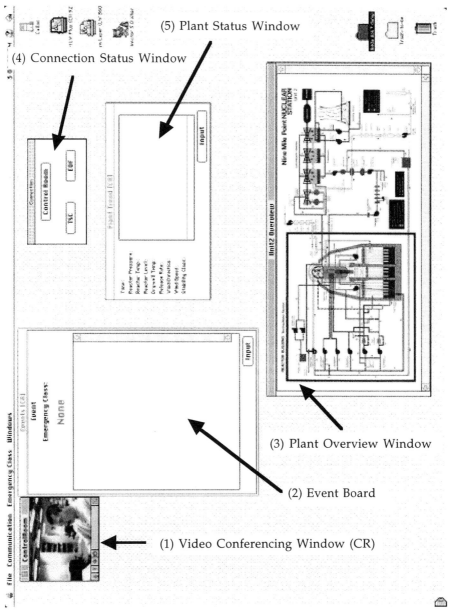

Figure VI.8: Prototype of multi-group ORM decision aid.

The system was customized for each site, so each system's functionality is based on the responsibilities of each group at that site.

The system functions are follows:

(1) Video Conferencing (all the sites).
(2) Talk, which enables a "chat" function between the sites (all sites).
(3) Event Board, which displays the status of the emergency including the declared emergency classes (all sites).
(4) Plant Status Board, which indicates the power plant's parameters input by an operators in CR (all sites).
(5) Connection Window, which displays the conditions of system connections between the sites (all sites).
(6) GIS Map, which indicates wind direction and speed and radioactive plumes. (TSC and EOF).
(7) Plant Overview, which displays the plant condition using graphical view (CR and TSC).
(8) Station Blackout Procedures, which indicate emergency procedures (CR and TSC).
(9) RPV Control Graphs, which are graphs that tracks plant conditions (CR and TSC).

B) Emergency Responses

When an emergency occurs, the System Shift Supervisor (SSS), who is the decision maker in the Control Room, reports the emergency to the plant authorities; then, TSC and EOF are staffed. When personnel of the TSC and EOF report in, the system connection is immediately established and data starts to be exchanged (Figure VI.9).

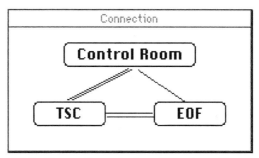

Figure VI.9: Connection window (Site Area Emergency).

Operational Risk Management

Assuming another RTE occurs that causes station blackout, the operators must conduct the appropriate emergency procedures in a correct and timely manner. The procedures are stored in the system for activation by the operator.

The geographic information system (GIS) shows the wind direction and speed that are obtained in CR and transferred to EOF in order to make recommendations for evacuation of the plant or surrounding community. Spreading radioactive plumes are calculated at TSC and the information is presented on the GIS display (Figure VI.10).

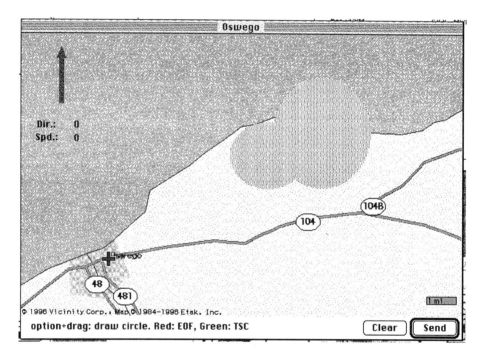

Figure VI.10: GIS around the power plant.

Reactor Pressure Valves (RPV) control graphs are displayed during the emergency operations (Figure VI.11). Eight graphs are used to monitor the plant conditions. The ERO tracks the plant parameters. All the data readings for these graphs are obtained in CR, and an operator clicks on the graph to transfer the data reading to TSC. Once the graph indicates that the plant parameters have entered an unsafe range, the personnel in TSC must recommend appropriate procedures for the operators.

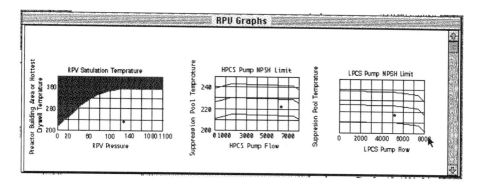

Figure VI.11: RPV control graphs.

Figure VI.12 shows the plant overview. The status of pumps and valves are displayed on a graphical view. If the ERO finds a pump that is malfunctioning, the information is transferred on this view by clicking on the image of the equipment. A cross emerges when the equipment is down.

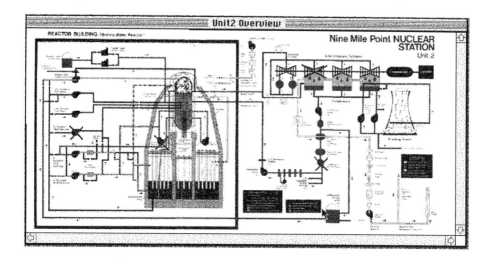

Figure VI.12: Plant overview.

C) Assessment

The prototype DSS was evaluated in two "gaming" situations, for two utilities that have nuclear power generation facilities. Officials and personnel from the State department that regulates these utilities assumed roles in the operation center. In addition, the technical support center and the emergency operations center also participated in an exercise that simulated an incident that resulted in a potential release of radioactive material. The three workstations supporting these facilities were connected as part of a local area network. Voice and video conferencing capabilities were provide in addition to the electronic transfer of data and multimedia displays. The hour long simulation enabled the participants to assess the technology and the operational risk management logic. The technology was supportive of the decision making process, although training in the use of the windows technology was required. The participants found the logic easy to use once they understood how to express their preferences – another training requirement.

3. Concluding Comments

The two applications discussed in this chapter show how the operational risk management logic can be used in cases where

- three-dimensional depiction of a physical setting is required;

- risk analysis is conducted in real-time to support command and control operations;

- the course of actions are sequences of activities and do not have a physical analogy like a roadway on a map; and

- multiple group are involved in the decision making process.

In addition, these applications capitalize upon advances in networking and multimedia technologies.

CHAPTER VII: ASSESSMENT OF OPERATIONAL RISK MANAGEMENT

1. Experimental Assessment of the Decision Logic

1.1 Introduction

The process of assessing safety (risks), efficiency (costs), and effectiveness (benefits) due to an RTE in an operational setting must be based on the cognitive capabilities and limitations of the human dispatcher. The decision models discussed in the previous chapters consider humans' limitations in assessment, a lack of complete and relevant information and knowledge, the potential for catastrophic events, and time constraints in the assessment and choice processes. These models also provide a framework for a consistent, flexible, and time-efficient assessment process. The subsequent choice process matches the assessment process by being intuitively sound and logical. In this chapter, an empirical investigation of appropriate real-time decision support for operational risk managers is provided for different models.

The experimental assessment was conducted with the prototype decision support system for transportation of hazardous materials which was discussed in Chapter IV. For this purpose, four alternative models for rerouting vehicles carrying hazardous materials are considered. The first two models support the dispatcher in making decisions by proposing alternative routes. One model uses an advanced visual interactive interface, where the dispatcher can construct alternatives to the planned routes using just his or her experience. The second model uses a simple heuristic that computes alternative routes and presents them to the dispatcher for choice.

The other two models support the dispatcher in making decisions based on the attributes of risk and cost (Chapter IV). One model is the ordinal preference routing model that is designed for the stressful environment of real time decision making, while the other uses a multiattribute routing approach such as those typically used for strategic route selection [List, et al., 1991]. The assessment of the four models, as embedded in a decision support system that employs multimedia technology, will be the focus of this chapter. The results of this experiment should provide the basis for designing routing models as part of real time tracking systems.

The section to follow will describe the decision problem and the four decision support models. Then, the hypotheses will be stated, followed by the design and implementation of the experiment. Finally, the results of the experiment will be presented together with concluding remarks.

1.2 The Decision Problem and Four Decision Models

New information and communications technology allows a dispatcher to monitor the hazardous material vehicles on a transportation network on a screen (e.g., on a PC) in real time. S/he gathers and processes data continuously about RTEs that could affect the safety and effectiveness of the shipments. The routes of some of the vehicles in the fleet will not be affected by the RTEs. Vehicles that are in transit and scheduled to pass through the area affected by the RTE may need to be either rerouted or delayed. Therefore, the dispatcher must address two tasks under time constraints and with incomplete information: (1) determine the affected vehicles and (2) evaluate alternative safe and cost-effective routes. Three rerouting decisions are possible: (1) to continue on the planned route, (2) to reroute to a safer and more cost-effective route, or (3) to stop at the next turnout.

Since a transportation network is an implicit representation of feasible routes for any origin/destination pair, the dispatcher's cognitive process for each rerouting decision consists of two stages. The dispatcher must first "construct" feasible routes and then choose the optimal one from among those routes. The dispatcher can use different strategies in this decision process. One is to start at the actual location of the vehicle and add links until the destination is reached. Another approach is to start from the planned route and delete and add links to construct the new route. Yet another approach is to look for intermediate nodes on the transportation network between the vehicle's actual location and its planned destination, and then to construct a route that connects these nodes in an appropriate sequence.

When the prototype decision support system discussed in Chapter IV was presented to experienced dispatchers, some argued that rerouting vehicles under time constraints can be done without support, while others claimed that such tasks quickly led to cognitive overload on the dispatcher, especially when the number of vehicles in a fleet is large. The experiment reported herein provides answers to how much decision support is reasonable for dispatchers using satellite tracking technology for controlling a truck fleet transporting hazardous materials.

One way to alleviate the cognitive effort is to let the system compute optimal routes. The dispatcher, however, must assess the effects of the RTEs on transportation safety and costs before optimal routes can be computed for all affected vehicles. The dispatcher must then decide whether to assign the vehicles to the optimal route, leave them on the planned route, or assign them to a completely different route. In order to use the computational approach, the dispatcher must reevaluate the risks and transportation costs for the portion of the transportation network affected by the RTE. This is decision making by attribute. If, on the other hand, the dispatcher is just presented with an electronic map display with the planned routes and the region that may be affected by the RTE (typical display of commercial satellite tracking systems), s/he will have to decide on rerouting by reviewing alternative routes. This is decision making by alternative.

Two of the four models compared in this chapter are based on decision making by **alternative** and the other two on decision making by **attribute**. The four decision models are compared with regard to their effort and accuracy, as proposed by Payne [1987], operationalized by Johnson and Payne [1985], and used to assess decision aids by Todd and Benbesat [1991]. Effort is measured as the time it takes the dispatcher to make rerouting decisions for all the affected vehicles, and accuracy is measured as the number of correctly rerouted vehicles given the impact of the RTEs. The four decision models are as follows:

Visual Interactive (VI) model (decision making by alternative): Whenever RTEs occur, only the planned routes of the vehicles are shown on the map on the screen as explicit alternatives. There is no other cognitive support for the construction of alternative routes or selection of the optimal route. If the dispatcher wants to choose a route different from the planned one, s/he can select and delete links by clicking on them with the computer mouse.

Conservative Heuristic (CH) model (decision making by alternative). Two alternatives are shown on the map on the screen: the planned route and the worst-case solution. The worst-case route is based on the heuristic that all areas affected by RTEs must be avoided, even if the RTE does not affect the risk. The dispatcher can choose one of the two routes or construct a different one, the same as in the VI model.

Ordinal Preference (OP) model (decision making by attribute): The two attributes used here for routing shipments of hazardous materials are risks and costs. Each entity, such as a bridge, a tunnel, a hospital, a school, a mountain road, or an inner-city road, is assigned a pair of risk and cost values. Risk values are "too high risk," "high risk," "low risk," and "no risk." Although safety has multiple dimensions, as noted by Kalelkar and Brooks [1978], categorization into four risk classes lets the dispatcher make his/her own assessment by, e.g., assigning situations that involve a threat to human life to "too high risk." Transportation costs were assumed to be 80 Swiss Francs (*SFr*) per hour. This means that driving with a velocity of 80 km/h results in costs of one *SFr* per km. The optimal route from the vehicles' present location to the planned destination is computed using the following priorities: (1) it never takes a link with an entity having "too high risk," (2) it avoids as much as possible links with entities having "high risks," (3) it minimizes transportation costs, and (4) it avoids as much as possible links with entities having "low risks."

Multiattribute Utility (MAU) model (decision making by attribute): The procedure of rerouting vehicles in cases of RTEs for the MAU model is identical to that of the OP model. The MAU model, however, uses numerical values for both risk and transportation costs as discussed in Chapter II, Section 2.4, and Chapter IV, Section 2.3. An increase in risk due to an RTE can be expressed on an cardinal scale from 0 to 100 where the dispatcher assigns a high value for situations that involve, for example, a threat to human life. When the impact of an RTE must be assessed, the risk and velocity values for normal driving conditions are shown on two scales. An increase in risk is assessed by moving the slide upwards. This new scale value $(x) \in [0,100]$ is then transformed to the absolute risk value $\lambda(x) \in [0,1]$ with the following formula:

$$\lambda(x) = \begin{cases} \bullet \ 10^{-[10-INT(x/10)]}, \text{ for } x \in Z = \{0,10,20,...,70,80,100\} \\ \bullet \ \left[x - 10 \times INT\left(\frac{x}{10}\right)\right] \times 10^{-[10-INT(x/10)]}, \text{ for } x \in [0,100] \setminus Z' \end{cases}$$

where for example, a scale value of $x=62$ results in an absolute risk value of $\lambda=2\times10^{-4}$. Changes in transportation costs are assessed in the same way as in the OP model. For each link, a total cost is computed by converting risk to a risk-cost value, converting velocity to transportation cost, and summing them. After all the affected entities have been assessed, the minimum cost routes are computed for all vehicles. The dispatcher can choose whether to take the new optimal route, to stay on the planned route, or to construct a different route by using the mouse as previously described.

When RTEs occur, the dispatcher must assess the impacts on risk and velocity of a vehicle driving through the affected area by changing the risk and the transportation costs. The increase in transportation costs that results from driving through the area affected by the RTE can be assessed by reducing the possible travel velocity within that area. The impact on safety of an RTE can be assessed by changing the risk value of the area affected by the RTE from "no risk" or "low risk" to "high risk" or "too high risk." After all the RTEs have been evaluated, the system computes new optimal routes for the affected vehicles and presents the solution, as well as the planned routes, to the dispatcher. If s/he decides not to choose either one, the dispatcher can construct a different route by using the mouse as in the VI or CH model.

Figure VII.1 shows the user interface for the OP (right) and the MAU model (left). The bar on top (timer) shows how much time is left for making the rerouting decisions of one scenario.

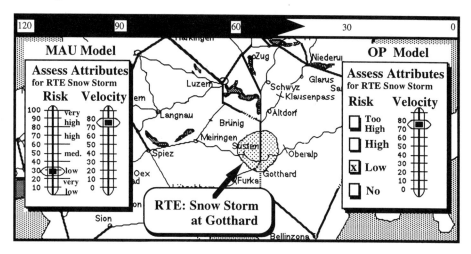

Figure VII.1. The user interface for the MAU (left) and OP (right) models.

1.3 Hypotheses

The Hypotheses were designed to examine effort and accuracy of the four decision models as imbedded into a decision support system. In summary, the four models are as follows:

<u>Visual Interface (VI) model</u>: Subjects see only the planned routes and the areas affected by the RTEs. No assessment of the attributes is possible and no heuristic for generation of additional routes is available.

<u>Conservative Heuristic (CH) model</u>: Subjects see the planned routes and for each vehicle the worst-case route that avoids the RTE. No assessment of the attributes is possible.

Ordinal Preference (OP) model: Assessment precedes the route selection. Risks are assessed on a ordinal preference scale, costs on a numerical scale. Subjects see the planned route and the route resulting from the assessment of the attributes for each vehicle.

Multiattribute Utility (MAU) model: Assessment precedes the route selection. Risk and costs are assessed on numerical scales. Subjects see the planned route and the route resulting from assessment of the attributes for each vehicle.

In the following hypotheses, effort refers to the total time required by the subjects to reroute the vehicles, while accuracy refers to the number of correct, acceptable or wrong reroutings. In this decision situation, the RTEs force the dispatcher to reassess risks and costs. However, not all RTEs affect the risk. A rerouting decision is considered correct if a vehicle is rerouted to avoid an RTE that increases the risk, or if a vehicle that plans to drive through an RTE that does not increase the risk is left on the planned route. A rerouting decision is considered wrong if the route passes through a region affected by an RTE that increases the risk, or if a vehicle is rerouted to avoid an RTE that does not increase the risk. Finally, it is considered acceptable if it is neither correct nor wrong; for example, it avoids a region affected by an RTE that increases the risk, but not on the fastest route.

A) Hypotheses Concerning Effort (HE)

HE 1: The VI model requires the greatest effort.
    Subjects using the VI model have only the choice between staying on the planned route and creating a new route. No support for creating routes is provided.

HE 2: Decision making by attribute (OP model and MAU model) requires less effort than decision making by alternative (VI model and CH model).
    Subjects using the OP model and the MAU model must spend time on the assessment of the attributes of risk and cost. However, an optimal route based on the assessment is computed and presented to the subjects automatically. The time invested in the assessment should be less than the time to create new routes.

HE 3: The OP model requires the least effort.
    In addition to HE 2, the OP model is expected to require less effort than the MAU model, since only ordinal measurement is used for risk.

HE 4: Effort for attribute assessment is less with the OP model than with the MAU model.

Subjects using the OP model have only the three choices: leaving the risk-preference constant, increasing it to high-risk, or closing the area affected by the RTE. Subjects using the MAU model can increase the risk level on a numerical scale from the actual risk level up to 100. It is assumed that it takes more cognitive effort to place the slide at a satisfactory position than to select one of three boxes. Costs are assessed for both models in the same way.

B) Hypotheses Concerning Accuracy (HA)

HA 1: The CH model produces the most wrong decisions.

It is expected that in such decision situations the dispatcher will select the new route based on the worst-case heuristic. This will result in reroutings that are not necessary.

HA 2: Decision making by attribute (OP model and MAU model) produces fewer wrong decisions than decision making by alternative (VI model and CH model).

It is expected that assessing the impacts of RTEs (OP model and MAU model) yields better rerouting decisions than choosing new routes (VI model and CH model) without support or with a simple heuristic (worst-case). This implies that decision making by attribute is superior to decision making by alternative in a real time environment with implicit alternative representation (i.e., the feasible routes are not shown explicitly on the map but must be constructed).

HA 3: The OP model is as accurate as the MAU model considering correct and acceptable decisions.

In addition to HA 2, it is expected that the OP model and the MAU model will not be significantly different in accuracy for these simplified scenarios. The differences between these models lie more in the details that become important when using this approach in a large-scale environment where a dispatcher is managing many vehicles and where the areas affected by RTEs are very large.

HA 4: The OP model is the most accurate model considering correct decisions.

The OP model is expected to perform best. This requires, however, that the assessment of the attributes be more accurate with the OP model than with the MAU model.

1.4 The Experiment

A) Data for the Task

A transportation network from Switzerland was chosen for this experiment. The first step was to assess risks and costs using technical reports from the Swiss Department of Transportation [BAP, 1988] and the literature [Glickman, 1991]; [Saccomanno and Chan, 1985] for normal driving conditions (i.e., no RTE is present). Planned routes could then be computed between any two origins and destinations on the network.

Two different measures were used to describe the risks: an ordinal and a numerical measure. The ordinal assessment was done by assigning highways a "no risk," and all other roads a "low risk" value. In addition, road segments with narrow curves, bridges, and/or tunnels; mountain roads; road segments with high average daily traffic; road segments through highly populated areas; and road segments through sensitive environmental areas were assigned "low risks." Under normal conditions, all entities were designated "no risk" or "low risk." Therefore, "high risk" situations could occur only during RTEs.

The numerical risk assessment was done for six different road types using accident data [BAP, 1988]; [Glickman, 1991]; [Saccomanno and Chan, 1985]. The six road types were: highways (at least two lanes with no cross traffic) in urban areas, highways in rural areas, canton roads (one-lane roads with cross traffic) in urban areas, canton roads in rural areas, mountain roads, and tunnels.

Transportation costs for normal driving conditions were computed by dividing all roads into four velocity classes: 50 km/h (mountain roads), 60 km/h (canton roads), 70 km/h (canton roads with partial highway), and 80 km/h (highways). With transportation costs being 80 *SFr* per hour, driving along a road segment of length L with velocity V results in costs of: $L \times (80/V)$ *SFr* (see also Chapter IV, Section 2.3).

For the MAU model, the numerical risk values were converted into costs and added to the transportation costs to arrive at a total risk-cost value. The amount society is willing to pay to save a life was estimated to be 10^7 *SFr*, based upon research by Fischhoff et al. [1981]. Under normal conditions all roads on the chosen network in Switzerland are rather safe. Thus, the "optimal" route between any two points under normal conditions is identical with the shortest route. For both the OP model and the MAU model, the planned routes under normal conditions are identical. These optimal routes under normal conditions were also used for the other two models.

B) Three Scenarios

The experiment was developed in a multimedia environment on a Macintosh II si in Hypercard and Pascal (Figure VII.2). The subjects used the mouse as the input device, and voice, text, and graphics as output in the human-machine interface. At the beginning of each of the scenarios, the origins, destinations, and planned routes of three vehicles were introduced to the subjects visually and by voice. After this introduction, the subjects used the screen to observe the vehicles move along their planned routes for about 10 seconds.

The experiment consisted of three different scenarios requiring rerouting decisions by the subjects. These scenarios were presented sequentially, with rerouting decisions made for each scenario. The scenarios were presented to all subjects in the same order. A scenario consisted of two independent, simultaneously occurring RTEs and three vehicles moving on planned routes. These RTEs were designed to require rerouting decisions by the subjects. One event was designed to increase the risk of the shipments while the other did not affect the risk. Two different areas in the region were affected by the RTEs, but none of the three vehicles was in the affected areas when the RTEs occurred. The vehicles were routed such that one vehicle is scheduled to go through the region affected by the RTE that causes an increase in risk, one through the region affected by an RTE that does not increase risk, and the third through both regions. There are three vehicles and three scenarios yielding nine decisions for each subject. Since all decisions are considered to be of equal value, the highest score that could be obtained by any subject was nine correct decisions. A possible score by a subject could be: three correct, four acceptable, and two wrong decisions (the sum of correct, acceptable and wrong decisions is nine). In addition, the computer measured the total time, and, for the OP model and the MAU model, the time for assessment.

For each RTE, a picture appears with the voice describing the event. Additional information is provided to the subjects in the form of text and graphics. This multimedia environment simulates a dispatcher's workstation and provides a realistic decision environment. The RTEs were taken from news accounts describing commonly occurring situations in the region, such as snow storms and traffic accidents.

Following the RTE alarm, a timer is activated and portrayed by a bar at the top of the screen that moves from left to right. When the bar reaches the right corner of the screen, 120 seconds have elapsed. This is the maximum time the subjects had to reroute the three vehicles for each scenario. The pilot test showed that experienced dispatchers could complete the task in this time period but did feel pressure. The pictures, voice alarms, and the time pressure were used to simulate the stress of crisis situations [Belardo et al., 1984], [Smart and Vertinsky, 1980], [Janis and Mann, 1977].

The procedure of the experiment for the four models is illustrated in Figure VII.2.

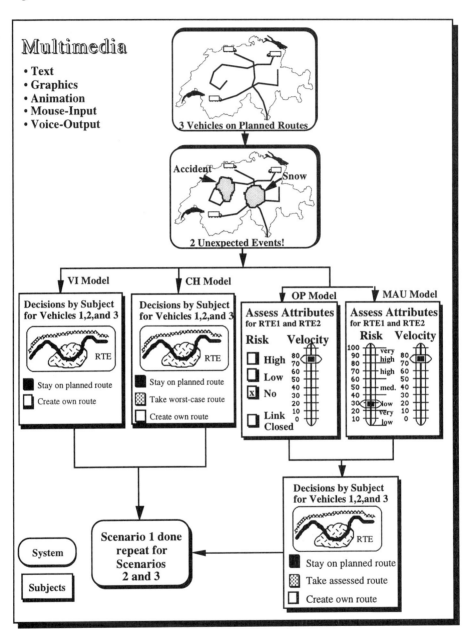

Figure VII.2. Procedure of the Experiment for the Four Decision Models.

- Subjects using the VI model see on the map the planned route and the two areas affected by the RTEs. They must first decide if they want the vehicle to continue on the planned route or to use another route. If the latter is the case, they can create a new route by clicking the mouse on the appropriate links. This rerouting process is repeated for all three vehicles.

- Subjects using the CH model can choose between the planned route and the worst-case route (i.e., the optimal route avoiding both affected areas), or create a new route, in the same way as did subjects using the VI model.

- Subjects using the OP model and the MAU model must first assess the impacts on risk and velocity within the affected areas. Subjects using the OP model change the risk level by clicking on the appropriate box, while subjects using the MAU model move a slide. The speed at which the vehicles can drive within the affected areas is changed in both the OP model and the MAU model by moving a slide (see Figure VII.1). After subjects using the OP model and the MAU model have finished the assessment, the optimal route is computed automatically for each vehicle. The planned route and the new computed route are presented to the subjects, who can select one of them or create a new route. A scenario is completed when routing decisions are made for all three vehicles.

C) Subjects

The experiment was conducted at the truck dispatching school in Wil, Switzerland. Thirty-two subjects participated in the experiment. They all had several years of experience as dispatchers and truck drivers. At the time of the experiment, they already had completed classes in Hazardous Material Transportation and Informatics.

The procedure of the experiment is illustrated in Figure VII.3. The thirty-two subjects were divided into two groups of sixteen. One group participated in the experiment in the morning, the other in the afternoon. Each group received a 45 minute introduction to general aspects of the experiment and basics of satellite tracking systems. Subjects were told that the nine routing decisions in the experiment had to be made as quickly and as accurately as possible. The subjects also were told that prizes would be awarded for the three best results.

After the introduction, the sixteen subjects were randomly divided into four groups of four for the four models. Four rooms, one for each model, were set up as dispatching center and used as experiment rooms. Each subject spent forty-five minutes in the corresponding experiment room. Four subjects, one in each room, were simultaneously performing the experiment; the other 12 subjects were waiting in a waiting room where videos about satellite tracking systems were shown. Subjects still waiting to perform the experiment could not talk to those coming back from the experiment. The session in the experiment rooms started with a short introduction to the experiment by the instructor.

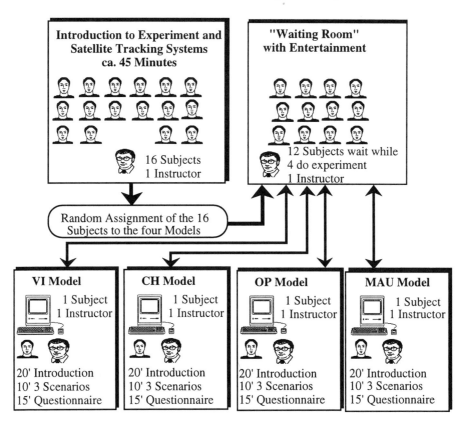

Figure VII.3: Procedure of the Experiment with one Group of 16 Subjects.

Subjects were told that they had to evaluate the routes of nine vehicles due to changes in the environment. The VI and CH subjects were told that whenever RTEs occur, they first must assess the impact on transportation risk and cost for all vehicles. Subjects using the OP and MAU models were instructed in using the risk and velocity scales. The definitions of correct, acceptable, and wrong routing decisions were known to all subjects. CH, OP, and MAU subjects were told that they could choose among the planned route, the newly generated route, or a self-generated route (i.e., they should be very critical of the routes generated by the system). The objective of the experiment was for the subjects to make as many correct routing decisions as rapidly as possible.

Then, the subjects could run an exercise scenario several times to get accustomed to the system and the procedure of the scenario. The experiment started after they felt comfortable with the use of the mouse and the procedure of the scenario. The subjects received a short break of about 15 seconds between scenarios. After completion of the experiment, subjects answered questions on the computer concerning their experience in dispatching, truck driving, and computing. Five questions were asked to ascertain their knowledge of the region, and 12 questions to assess the ease of use of the system and the support provided for risk assessment and decision making. Then, the four subjects went back to the waiting room and four new subjects started the sessions in the experiment rooms. This procedure was repeated with sixteen different students in the afternoon.

1.5 Results

A) Overview

Table VII.1 shows the results of the experiment. The vehicles that plan to drive through the area affected by the RTE that does not increase risk must not be rerouted (Vehicle 1/Scenario 1, V2/S2, V3/S3). The routing decision for one of these vehicles is correct if it stays on the planned route, acceptable if it still passes trough the affected area but not along the fastest route, and wrong if it avoids the affected area. The vehicles that plan to drive through the area affected by the RTE that increases the risk must be rerouted (V3/S1, V1/S2, V2/S3). The routing decision for one of these vehicles is correct if it avoids the affected area on the fastest route, acceptable if it avoids the affected area but not on the fastest route, and wrong if it still goes through the affected area. The vehicles that plan to drive through the areas affected by both RTEs must also be rerouted (V2/S1, V3/S2, V1/S3). The routing decision for one of these vehicles is correct if it avoids the area affected by the RTE that increases the risk on the fastest route (the fastest route would still go through the area affected by the other RTE), wrong if it passes trough the area affected by the RTE that increases the risk and/or if it avoids the area affected by the RTE that does not increase the risk, and acceptable otherwise.

Since eight subjects worked with each model, the sum of correct, acceptable, and wrong solutions for each vehicle and scenario is eight. The sum of the scores for each model is 72, since rerouting decisions must be made for nine vehicles.

Vehicles:	Scenario 1			Scenario 2			Scenario 3			= 72	
	V1	V2	V3	V1	V2	V3	V1	V2	V3	Total	p,c,w/a
Vehicle is affected by:	not haz. RTE	both RTEs	haz. RTE	haz. RTE	not haz. RTE	both RTEs	both RTEs	haz. RTE	not haz. RTE		
VI Model Scores	7C,p 1A,c 0W	6C,c 2A,c 0W	8C,c 0A 0W	7C,c 0A 1W,s	4C,p 0A 4W,c	3C,c 3A,c 2W,c	5C,c 3A,c 0W	2C,c 6A,c 0W	8C,p 0A 0W	50C 15A 7W	19,31,0 0,15,0 0,7,0
Times	84.4±23.8			108.6±15.3			91.9±17.0			94.7±21.0	
CH Model Scores	8C,p 0A 0W	3C,c 0A 1W,p/4W,w	8C,w 0A 0W	8C,w 0A 0W	8C,p 0A 0W	1C,c 1A,c 6W,w	4C,c 1A,c 2W,p/1W,w	6C,w 1A,c 1W,p	8C,p 0A 0W	54C 3A 15W	24,8,22 0,3,0 4,0,11
Times	52.2±28.0			52.3±28.0			63.0±37.5			54.2±30.5	
OP Model Scores	8C,p 0A 0W	8C,a 0A 0W	8C,a 0A 0W	8C,a 0A 0W	8C,p 0A 0W	8C,a 0A 0W	7C,a 0A 1W,p	5C,a 2A,c 1W,p	7C,p 0A 1W,a	67C 2A 3W	23,0,44 0,2,0 2,0,1
Times	42.8±5.0*			43.3±12.1*			53.3±22.2*			46.4±14.4*	
MAU Model Scores	5C,p 0A 3W,a	3C,a 1A,c 3W,a/1W,c	8C,a 0A 0W	7C,a 0A 1W,p	6C,p 0A 2W,a	4C,a 0A 4W,a	5C,a 2A,c 1W,p	7C,a 0A 1W,p	6C,p 0A 2W,a	51C 3A 18W	17,0,34 0,3,0 3,1,14
Times	61.6±19.3			56.6±10.3			51.4±21.1			56.5±17.3	

C: correct solution
A: acceptable solution
W: wrong solution

haz. RTE: RTE increases risk
not haz. RTE: RTE does not affect risk

p: planned route has been chosen
c: route has been created
w: worst-case route has been chosen
a: result of assessment has been chosen

Times: Average time of the 8 subjects to complete 1 scenario (3 routing decisions) ± standard error.
*: Four of the 8 assessment times were lost due to computer error. Thus, only 4 subjects are considered.

Table VII.1: Results of experiment.

Time is recorded commencing with the RTE alarm. Total time for rerouting the three vehicles is recorded for all models. In addition, the times for assessment of risk and velocity in the OP and MAU model were also recorded. An example of how to read Table VII.1 is given for the CH model, V1/S3: since the vehicle is affected by both the RTEs, it must be rerouted; four of the eight subjects created a correct solution (4C,c), one subject created an acceptable solution (1A,c), two subjects stayed on the planned route which is wrong (2W,p), and one subject picked the worst-case route which is also wrong (1W,w).

According to the design of the experiment and the characteristics of the four models, subjects working with the VI model could have obtained, with no effort, three correct solutions by staying on the planned route when it was appropriate to do so. To obtain the other six correct solutions, the subjects had to create their own routes. Subjects working with the CH model could have obtained, with no effort, six correct solutions, either by staying on the planned route for the vehicles that were affected only by the RTEs that did not increase risk, or by taking the worst-case route for the vehicles that were affected only by the RTEs that did increase the risk. To obtain the other three correct solutions, the subjects had to create their own routes. Subjects working with the OP and MAU models first had to assess costs and risks and then had to make a decision. In three cases, the correct solution was to stay on the planned route (for the vehicles that were only affected by the RTE that did not increase the risk); for these cases, no assessment would have been required. After the assessment was done, insignificant effort was required to make the decisions.

B) Results Concerning Effort

Table VII.2 shows the time it took the 32 subjects to perform the experiment. The maximum time given to each subject to perform one scenario was 120 seconds (360 seconds for the three scenarios). Table VII.3 gives the results of the pair-wise testing of the models. Since subjects working with the CH, OP, and MAU models needed significantly less time to fulfill the tasks than subjects working with the VI model, the hypothesis HE 1 is accepted at a level of significance of 5% (see [Conover, 1980] for procedure). However, HE 2, that decision making by attribute (OP and MAU models) requires less effort than decision making by alternative (VI and CH models), cannot be concluded. The hypothesis that the OP model and any of the other models require the same effort has been rejected only for the VI model but not for the CH and MAU models. Therefore, HE 3 cannot be concluded. However, it must be noted that CH subjects solved their tasks rapidly but obtained a rather low score.

	Subject #									μ	σ
	1	2	3	4	5	6	7	8			
VI Model	232	288	244	324	276	296	338	281	2279	284.9	36.0
CH Model	64	136	131	211	151	265	117	226	1301	162.6	66.0
OP Model	190	137	119	111	*	*	*	*	557	139.3	35.5
MAU Model	156	117	211	143	186	202	210	132	1357	169.6	37.3

*: Due to computer error, four assessment times were lost.

Table VII.2: Time to reroute 3 vehicles for three scenarios (9 decisions).

	VI	CH	OP	MAU	
		→ 2.1%	→ 0.7%	→ 0.08%	VI
			↔	↔	CH
CH OP ↘ ↙ VI ← MAU				↔	OP
					MAU

Ho: Model X and Model Y have the same accuracy
H1: The accuracy of the two models is not equal

Mann-Whitney Test:

X→Y: Model X is at level $\hat{\alpha}$ (%) more accurate than Model Y.

↔: Ho is not rejected at level $\hat{\alpha}$ 10%.

Table VII.3: Comparison of effort (time).

Table VII.4 shows the times for the 16 subjects to assess the attributes of risk and velocity. Due to computer error, data on four subjects were lost for the OP model. Nevertheless, HE 4 is accepted at a level of significance of 10% (critical level=7.3%). It can be further concluded that the OP model is faster for assessment than the MAU model.

	Subject #								μ	σ
	1	2	3	4	5	6	7	8		
OP Model	79	55	63	67	*	*	*	*	66.0	10.0
MAU Model	79	69	80	76	96	92	91	61	80.5	12.1

* Due to computer error, four assessments for the OP model were lost.

Table VII.4: Time for OP and MAU subjects for assessments of risks and velocities.

C) Results Concerning Accuracy

Table VII.5 shows the scores of the 32 subjects; i.e., the correct, acceptable, and wrong solutions. It can be seen that the VI model has a high number of acceptable solutions (15); i.e., routes that are safe but not efficient. The CH and the MAU models have a high number of wrong solutions (15 and 18, respectively), while the OP model has the highest number of correct solutions (67).

	Subject #								
	1	2	3	4	5	6	7	8	
VI Model	4/4/1*	7/1/1	8/1/0	4/4/1	8/1/0	7/0/2	6/1/2	6/3/0	50/15/7
CH Model	6/0/3	6/0/3	7/0/2	8/0/1	7/0/2	7/2/0	7/0/2	6/1/2	54/3/15
OP Model	8/1/0	8/1/0	9/0/0	9/0/0	9/0/0	9/0/0	8/0/1	7/0/2	67/2/3
MAU Model	8/0/1	6/0/3	8/0/1	5/0/4	7/1/1	5/1/3	6/1/2	6/0/3	51/3/18

*: Subject 1 of the VI Model had 4 correct, 4 acceptable, and 1 wrong solutions.

Table VII.5: Score per subject and model (each model had different subjects).

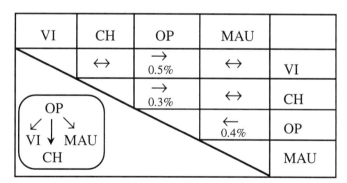

Ho: Model X and Model Y have the same accuracy
H1: The accuracy of the two models is not equal
Mann-Whitney Test:
X→Y: Model X is at level $\hat{\alpha}$ (%) more accurate than Model Y.
↔: Ho is not rejected at level $\hat{\alpha}$ 10%.

Table VII.6: Accuracy as score of subjects per model (for correct solutions).

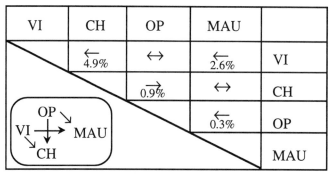

Ho: Model X and Model Y have the same accuracy
H1: The accuracy of the two models is not equal

Mann-Whitney Test,

X→Y: Model X is at level $\hat{\alpha}$ (%) more accurate than Model Y.
↔: Ho is not rejected at level α 10%.

Table VII.7: Accuracy as score of subjects per model for (wrong solutions).

Table VII.6 and VII.7 give the results of the statistical tests. It shows that HA 1, which states that the CH model produces the most wrong solutions, is accepted at a 10% level of significance only for the OP and VI models but not for the MAU model (Table VII.7). It is interesting to note that subjects working with the MAU and the CH model made most of the wrong rerouting decisions because they chose the new routes proposed by the computer. For the MAU model, 14 of the 18 wrong solutions were obtained by choosing routes computed by the system (based on the subject's risk/velocity assessment). For the CH model, 11 of the 15 wrong solutions were obtained by choosing the worst-case solution when it was not appropriate to do so (see Table VII.1).

HA 2, which states that decision making by attribute (OP and MAU models) is more accurate than decision making by alternative (VI and CH models), holds only for the OP model. This implies that decision making by attribute is only superior to decision making by alternative if the assessment process is supported appropriately.

HA 3, which states that the OP model and the MAU model have the same accuracy, has been rejected (Table VII.6 and VI.7). Thus, the OP model is superior to the MAU model. Using a set of predefined risk classes instead of a numerical scale seems to have a bigger impact than expected.

Table VII.6 shows that the OP model is the most accurate, at a level of significance of 1% (critical level =0.5%). Therefore, HA 4 is accepted.

These results imply that decision making by attribute can be quite superior to decision making by alternative. This statement cannot be generalized, however, since the results depend heavily on the support for risk/cost assessment.

D) Effort and Accuracy

Accuracy is measured as the number of correct decisions. A rerouting decision is considered "correct" if the route is both safest and minimum cost (i.e., fastest); it is considered "acceptable" if the route avoids a hazardous RTE but is not the fastest one, otherwise it is considered "wrong." Since every subject had to make nine rerouting decisions, the sum of "correct," "acceptable," and "wrong" decisions is nine. For example, a possible score is five correct, two acceptable, and two wrong decisions (2c/5a/2w), for a total of nine decisions. Figure VII.4 shows the relation between accuracy (scores) and effort (decision time) for the eight subjects per model, where score is defined as the sum of correct decisions plus acceptable decisions.

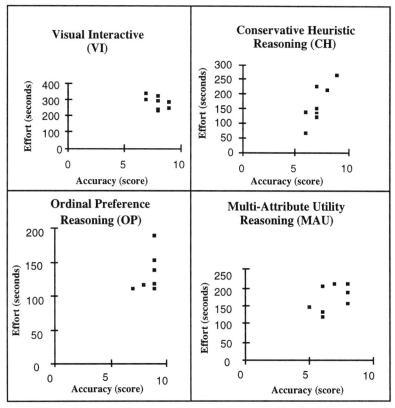

Figure VII.4: Accuracy (score: correct plus acceptable decisions) vs. effort (decision time) for the four reasoning approaches.

The CH and MAU models show a positive correlation between effort and accuracy (Figure VII.4); i.e., the more time the subjects invested in decision making, the higher the score. The VI model shows a negative correlation. This stems from the fact that constructing new routes by clicking with the computer mouse on the appropriate road segments yields higher scores (accuracy) but also takes more time (effort). The relation between effort and accuracy for the OP model seems to be different. Low scores were obtained where decisions were made rapidly. The best scores (i.e., all nine solutions are correct) were obtained if subjects invested more than 120 seconds (maximum time available for each subject is 360 seconds). These results are statistically significant at the 5% level for the CH model, and at the 10% level for the VI model (omitting one outlier), while the results for the OP and MAU models represent merely trends.

E) Task Complexity and Accuracy

Task complexity can be defined in different ways. Since the number of vehicles to be rerouted and the number of RTEs were constant for all three scenarios, task complexity can be defined as the type of RTE (hazardous or harmless) and number of RTEs that affected the vehicles. Of the two RTEs that occurred in each scenario, one did increase the risk (hazardous) and the other did not (harmless). For every scenario, one vehicle was affected by the hazardous RTE, another by the harmless RTE, and the third by both RTEs. The vehicles that were affected by the corresponding RTEs were determined randomly and were the same for all subjects. The most complex rerouting decisions of the three, is the one where the vehicle is affected by both RTEs. For the other two decisions, task complexity is assumed to be higher when effort (time) to assess the RTE is higher. The OP subjects took 11.2±3.5 seconds to assess the harmless RTE and 10.8±2.9 seconds to assess the hazardous RTE. The MAU subjects took 12.1±2.9 seconds to assess the harmless RTE and 15.2±4.6 seconds to assess the hazardous RTE. Since there is no significant difference in the RTE assessment times for both models, assessing a harmless RTE and assessing a hazardous RTE both have the same task complexity.

The effort to choose a route for the three vehicles were not recorded; only the overall choice times for all three vehicles per scenario were recorded. The choice times per vehicle could have given insights about the relation between task complexity and effort (e.g., more complex tasks might require more effort to solve).

Table VII.8 shows the relationship between accuracy (scores for correct/acceptable/wrong rerouting decisions) and task complexity. It can be seen that accuracy is lowest when vehicles were affected by both RTEs. This result holds for all models except for the OP model. Accuracy for the OP model seems to be independent of task complexity. When only "correct" solutions are considered, the VI and the CH models also show a decrease in accuracy between hazardous and harmless RTE. Accuracy with the VI, CH, and MAU models decreases with increasing task complexity, while again the OP model seems not to be affected by increased task complexity.

	Scores for vehicles that were affected by:		
Model	harmless RTE	hazardous RTE	both RTEs
VI	19/1/4	17/6/1	14/8/2
CH	24/0/0	22/1/1	8/2/14
OP	23/0/1	21/2/1	23/0/1
MAU	17/0/7	22/0/2	12/3/9

Table VII.8: Accuracy (correct/acceptable/wrong decisions) and task complexity (harmless, hazardous, and both RTEs).

It is interesting to note that CH subjects had a very low score for vehicles that were affected by both RTEs. One explanation is that both the current route and the new route were wrong for the vehicles that were affected by both RTEs. For the vehicle that is affected only by the harmless RTE, the correct route is identical with the original route. For the third vehicle, which is affected only by the hazardous RTE, the correct route is the one computed with the conservative heuristic.

The daily work of operators involves higher task complexity and consequently more time pressure than simulated in this experiment. For example, a dispatcher of vehicles carrying hazardous materials must handle more vehicles, and RTEs occur more frequently. The most appropriate reasoning approach seems to be the OP model. It is based on a lexicographic reasoning approach which (i) provides more support than the VI model, (ii) is more flexible than the CH model, and (iii) is more conservative than the MAU model. In fact, it has been found that decision makers are less likely to take risks (more conservative) under time pressure [Ben Zur and Breznitz, 1981].

F) Cognitive Load

Cognitive load can be characterized by the effort expanded on assessment and choice. Table VII.9 shows the efforts (times in seconds) for assessment and choice for the CH, OP, and MAU reasoning logics. Given are the average times (efforts) for decision making for those scenarios where the subjects made no mistakes ("no-mistake scenarios; i.e., both assessments and choices are correct), and the average times (efforts) of the other scenarios (i.e., some mistakes were made). Note that VI subjects could not assess the RTEs and had no new routes proposed by the decision support system. The assessment times for the CH model are zero, since the assessments were done automatically with a conservative heuristic.

		CH	OP	MAU
Effort for Assessment	scenarios without mistakes	0	22.3±4.2	28.1±7.1
	scenarios with some mistakes	0	21.0±7.2	26.5±5.6
Effort for Choice	scenarios without mistakes	82.6±14.0	17.3±3.5	26.8±7.0
	scenarios with some mistakes	40.3±27.0	41.3±20.0	30.2±18.2

Table VII.9: Cognitive load with OP and MAU models measured in seconds.

The efforts for the "no-mistake" scenarios can be used to assess the effects of the reasoning logic on correct solutions. The differences between the OP and MAU models for both the average assessment times and the variances are not significant. Thus, there is no benefit in using either reasoning logic in regard to the assessment-effort. On the contrary, the dispatcher must spend more time when using the OP and MAU models as compared to the CH model, which does the assessment automatically.

For the choice phase, the CH model is counter productive. Choice-effort in "no-mistake" scenarios is significantly greater than in the other scenarios. A possible reason for this is that CH subjects realized that many routes proposed by the conservative heuristic were wrong and they spent time determining the correct route.

The variances for choice-times are for all three reasoning approaches significantly smaller for "no-mistake" scenarios than for the other scenarios. This means that the effort expended on wrong choices varied more than the effort expended spent on correct choices. A possible explanation for this is that wrong choices were obtained either by rapid, random selection of one of the two routes proposed by the system, or by spending time fruitlessly in searching for the correct solution.

Two of the three consecutive tasks that were part of this decision process are assessment followed by choice (as shown in Figure VII.2). The OP model significantly reduces the choice-effort, while the MAU model does not. Moreover, although the OP and MAU models do not have significantly different choice-efforts, the OP model has significantly smaller variances in choice-effort than the MAU model. Therefore, the effort expended on correct decisions ("no-mistake" scenarios) is more stable, i.e., the variances in effort are smaller than for when mistakes are made.

High accuracy (high number of correct decisions) and stable effort (small variance in the effort to obtain the high accuracy) are related to each other; that is, high accuracy is not completely random (at least less random than low accuracy). There seems to be an optimal amount of effort needed to obtain maximum accuracy; more or less than this effort results in inferior solutions.

From a design perspective, a reasoning logic should not have as one of its goals rapid decision making but should rather strive for an appropriate balance between amount of decision support and task complexity. Since tasks in real time decision making may have different task complexities, different decision support approaches should be provided. This could mean that a decision support system should provide different assessment and choice strategies. In fact, cognitive load can often be reduced more effectively by switching choice strategy than by speeding up the current strategy [Hulland and Kleinmuntz, 1994].

G) Final Judgment

After the dispatchers of the OP and MAU models assessed the impact of the RTEs on risk and costs, new routes were computed automatically by the system and presented to them for selection. For the CH model, subjects had nothing to assess since new routes were computed automatically using a conservative heuristic. Therefore, the picture on the screen for the choice phase (see Figure VII.2) was the same for the CH, OP, and MAU models. This picture contained the current routes as well as the newly computed routes, all of which were presented for selection. In addition, subjects could decide to construct a completely different route if they decided not to take either the current or the new route.

For the choice process (i.e., the selection of new routes for the three vehicles), CH, OP, and MAU subjects were presented with two routes (the original and the new one) to choose from for every vehicle (one at a time). Some of the subjects carefully examined these two routes in the context of the geographic display. Less critical subjects would trust the new routes proposed by the system. In other words, some CH subjects accepted the routes generated by the conservative heuristic, and some OP and MAU subjects simply followed their assessments.

Figure VII.5 shows the conditional scores of correct plus acceptable solutions versus wrong solutions per model, given either that the screen presented the correct route (solution) or that neither route on the screen was correct. The correct solution was either the new computed route (when the vehicle was affected by a hazardous RTE) or the original route (when the vehicle was affected by a harmless RTE).

OP subjects were presented the correct solution on the screen (either the original route or the new computed route based on their assessment) in 89% of the choice situations. If they saw the correct route, they never made a "wrong" choice. CH and MAU subjects were shown the correct route 67% and 68% of the time, respectively.

Of special interest are those cases, where the subjects were not shown the correct route on the screen. For these cases, OP subjects selected the correct solution more often (63%) than the CH (42%) and MAU (35%) subjects. It is rather surprisingly that MAU subjects had the lowest frequency for "savings" (35%); that is, making the right choice when only incorrect solutions were presented for choice.

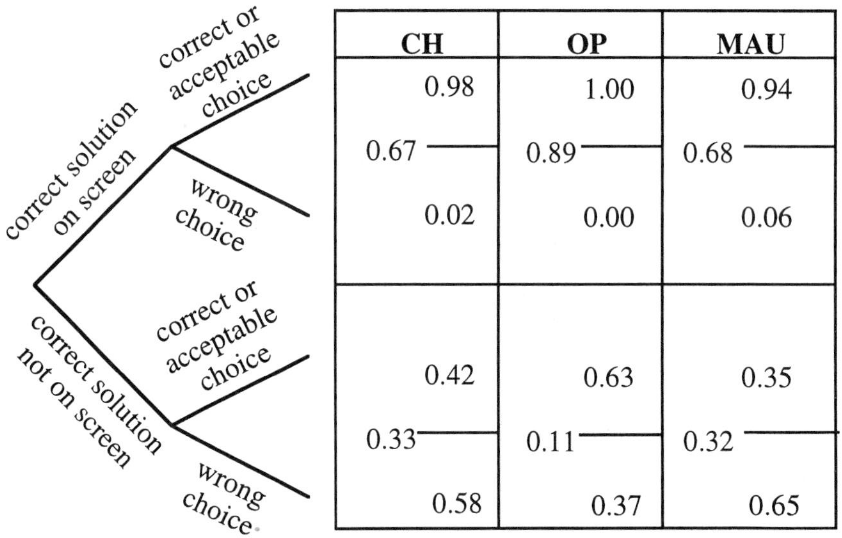

	CH	OP	MAU
correct or acceptable choice	0.98	1.00	0.94
correct solution on screen	0.67	0.89	0.68
wrong choice	0.02	0.00	0.06
correct or acceptable choice	0.42	0.63	0.35
correct solution not on screen	0.33	0.11	0.32
wrong choice	0.58	0.37	0.65

Figure VII.5: Marginal frequencies for seeing "correct" solution on the screen, and conditional frequencies for choosing the "wrong" solution.

H) Additional Results

It can be seen from Table VII.1 that subjects using the VI model did create significantly more routes (53=31+15+7, in the last column) than the other subjects (CH: 11, OP: 2, and MAU: 4) since they did not have any decision support. In addition, it was expected that subjects doing decision making by attribute (OP and MAU models), would create fewer routes for each vehicle than subjects using the CH model. However, the latter result is not significant.

In the final questionnaire, subjects using the OP model were asked if they would rather use a scale from 0 to 100 instead of the four classes for risk assessment. None of the subjects answered "for sure" or "possibly yes." Three answered "it does not matter," three "possibly no," and two "in no case." Thus, 100% of the subjects who worked with the OP model are not opposed to or even favor using this approach. Subjects using the MAU model were asked if they would rather use only the risk classes (as with the OP model) as opposed to numeric subdivisions of these classes (as with the MAU model). Only one subject answered "in no case," one chose "possibly no," two "it does not matter," three "possibly yes," and one "for sure." Therefore, 75% of the subjects who worked with the MAU model are not opposed to or even favor using an OP model. The MAU model not only performs worse than the OP model, it also was deemed less desirable by the subjects. Finally, evaluation of the questionnaires showed that computer experience did not affect the results of the experiment.

2. Simulation of Multi-Expert Decision Making

The issue of preference aggregation of multiple experts where knowledge is solicited to support the assessment of the impact of RTEs on a proposed activity or action is addressed first. Then, the process of deciding on a course of action is considered, using the assessments of multiple experts. The simulations use the ORM logic for multiple expert decision making (Chapter II, Section 3) and the data collected in the experiment discussed in the preceding section.

2.1 Preference Aggregation

Because experimental data for only six RTEs were collected, a resampling of the RTEs was necessary. Each RTE was repeated three times, resulting in a total of 18 RTEs. Figure VII.6 and VI.7 show the results of the aggregation of the velocity and risk for the MAU model over time. The κ-value was chosen to be ∞; that is, the weight of each expert depends only on the larger of the two values c_{ilk} (consistency) and d_{ilk} (deviation from aggregated mean). The band in Figure VII.6 and VI.7 is the resulting band of acceptable assessments for the most consistent group of experts. To compute this band (and the corresponding aggregated group assessment), the assessment values were rearranged in descending order. In addition, the weights of the experts were held fixed; that is, the assessment of each expert was considered exactly once.

Obviously, because of this high consistency, the assessments of all experts for the velocity assessment, and at least seven out of eight for the risk assessment were accepted for all RTEs. This explains why the band is rather large; that is, the double standard deviation of the assessments. The band of acceptable assessments depends on the consistency of the group of experts, the extreme values in relation to all assessments, and on the weights of the experts. If the weights w_{ilk} were not held fixed, but instead computed according to the formula proposed in the model (Section 3.1), the number of acceptable assessments would be smaller and thus the band also narrower.

For the other extreme case, where the experts are least consistent, the velocity values were rearranged alternatively in descending and ascending order, that is, the first expert was in the first assessment the most conservative, in the next assessment the least conservative, then again the most conservative, etc. Because of the low consistency, no assessment was ever acceptable for all RTEs; for such cases, the group assessment should be rejected. The reason for this situation is that the group consistency coefficient, C_k, was extremely low (close to 0), resulting in no band at all. However, the resulting group assessment was determined by a weighted mean of all eight assessments. Consequently, the aggregated group assessments of (i) the most consistent, (ii) least consistent, and (iii) non-weighted averages are almost the same (see Figure VII.6).

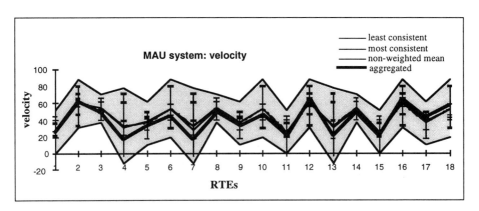

Figure VII.6: Aggregation of measured assessments (thick line) vs. aggregation of models (thin lines).

These three special cases show that taking the (possibly weighted) average can be misleading; that is, the assessment could be considered "best" (most consistent) or "worst" (least consistent). The proposed aggregation procedure discriminates for different experts' weights. For example, RTE4 in Figure VII.6 shows that the group assessment can be substantially lower that the simple average. The reason for this is that outliers are rejected iteratively. Thereby, an outlier is determined not only by the numeric value of the assessment but also by the expert's weight. Thus, what in numeric terms might not be an outlier, could very well be in terms of group assessment. For the MAU model, the average group consistency for the velocity assessment was $C_k = 0.43$, and for the risk assessment it was $C_k = 0.39$. For the OP model, the average group consistency for the velocity assessment was very low, $C_k = 0.14$. The reason for this is that if one assesses the risks with an ordinal scale, s/he often does not assess the velocity for high risk RTEs. In addition, if an RTE gets assigned an α risk preference, the velocity is automatically set to zero because the area affected by the RTE cannot be passed through.

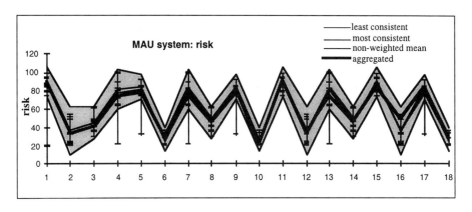

Figure VII.7: Aggregation of measured assessments (thick line) vs. aggregation of models (thin lines).

For the velocity assessments in the MAU and OP model, and the risk assessment in the MAU model, the aggregation of cardinal values has been done similar to the case discussed in Figure VII.6. The numeric risk values showed less consistency than the numeric velocity values for the MAU model.

The ordinal risks assessed in the OP model were at least sufficient; that is, a majority of the experts agreed in the tradeoff between risks and costs. For the assessment of low-risk RTEs, the assessment was always efficient; in fact the experts agreed unanimously on LR. For one high-risk situation, four experts recommended shutting down the affected area (α-value) and four left it open to traffic but assign an HR value. If there is only one RTE, then the reroutes are the same for α and HR assessments; however, if there are multiple RTEs or no other physically possible reroute, the resulting reroutes would differ. Thus, such an assessment is not feasible. In cases of sufficient but not efficient assessments, the group assessment is usually determined by the experts' weights. In the above example, the sum of the experts weights determined whether the area affected by the RTE must be avoided or not.

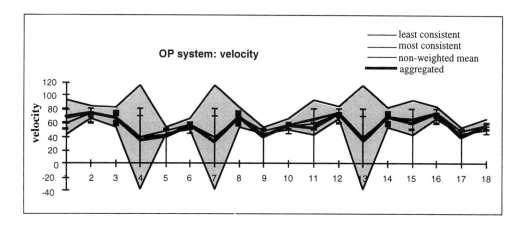

Figure VII.8: Aggregation of measured assessments (thick line) vs. aggregation of models (thin lines).

2.2 Choices

The CAs (routes) that resulted from the aggregated numerical values were all meaningful because a majority of the experts agreed on the CAs for all RTEs. Because of the recognition axiom, however, there is little meaning in debating about the quality of the reroutes resulting from the aggregated assessment. Moreover, reroutes are only suggestions both for the operational manager (dispatcher) and for the operator (driver). The operational manager always has the option of choosing his/her own CA by selecting or de-selecting activities of other CAs or simply by "constructing" his/her own CA.

In most cases, there are only two CAs based upon our assessment procedure – decision making by attribute – the planned CA and the CA which avoids the RTE. Consequently, changing a CA based upon the group assessment will result in one of these two alternatives. CAs based upon experts with acceptable assessments could differ from the group CA and from each other. If the choice based upon group assessment is the same as the choices made by the majority of the experts whose assessment was acceptable, this alternative should be selected. Both the planned CA and the alternative CA can be displayed graphically on the graph structure with the choice based upon the group assessment highlighted for the decision maker.

In the case of multiple RTEs, there may be more recommendations than just the planned CA and one alternative. There could be multiple recommended CAs – one based upon the group assessment and one based upon the acceptable assessments. The planned CA (if recommended or not), the new CA based upon the group assessment, and the new CA based on the acceptable assessments, could be presented to the decision maker. Since the number of experts is small, typically less than ten (in our example it was eight), displaying the foregoing alternatives should not present a problem for the decision maker. However, if there is a very complex decision situation with many vehicles and many RTEs – and there is no recommended new CA (including the planned CA) that is chosen by a majority of those experts with acceptable assessments – only the planned CAs and the CA chosen based upon the group assessments should be displayed. In addition, the CAs chosen based upon the individual experts whose assessments were acceptable should also be available; that is, stored in the computer for display.

3. Conclusions

An experimental assessment was made of four decision models as part of a decision support system for the monitoring and control of hazardous material shipments. They range from no reasoning support, to ordinal preference model, to numerical utility model, and finally to automatic assessment with a conservative heuristic. The task was to respond to a sudden onset event that could increase the risks and costs of hazardous material transportation. The decision support system developed for the test used multimedia technology to simulate the dispatcher's workstation, including electronic maps, scanned pictures and photos, graphical displays, voice alarms, and mouse input. The system was developed on a Macintosh II si in Hypercard and Pascal.

The four models were: (i) the visual interactive (VI) model; (ii) the conservative worst-case heuristic (CH) model (only safety of the shipments is considered, not the transportation costs); (iii) the ordinal preference assessment (OP) model (risks are assessed ordinally and costs numerically); and (iv) the multiattribute utility (MAU) model (risks and costs are assessed numerically).

The experiment was conducted at a school for truck dispatchers in Wil, Switzerland, with 32 experienced dispatchers and truck drivers. Each subject received a 45 minute introduction to the task and technology and the decision support system itself. The experimental procedure simulated the stress that would be evident in a realistic crisis situation by using voice alarms and a timing device.

The results show that decision making by attribute requires less effort and is more accurate than decision making by alternative if the assessment of safety and costs is based on a preference structure and not on a numerical scale. This is supported by the finding that 75% of the subjects working with numerical scales for safety assessment would like to work with ordinal classes, but no subject working with ordinal classes would prefer numerical scales. Moreover, heuristic and utility reasoning models have a positive correlation between effort and accuracy, while the VI model does not. The OP model seems to yield high accuracy independent of the decision effort. In addition, the results show that the accuracy of the OP model is not affected by task complexity, while the other models showed decreased accuracy with increasing task complexity. Comparing the three reasoning approaches that computed new routes after an RTE (CH, OP, and MAU model), the OP model creates less cognitive load than the other two models. In addition, the OP model reduces cognitive load more the than the MAU model, while the OP model can be counter productive. Finally, operators working with the CH and MAU models seem more easily to accept incorrect results generated by the system than those working with the OP model.

The limitations of a field experiment were kept to a minimum. The subjects, experienced dispatchers and truck drivers, confirmed the realism of the scenarios. The setting was such that each subject had a separate room for the experiment and a work space similar to a dispatcher's workstation. The technology was developed to simulate both the type of dispatcher's workstations presently on the market and the displays that could be developed using software currently available. Anecdotal evidence supported the existence of time pressure.

The experimentation employed as subjects individuals whose daily work involves dispatching trucks carrying hazardous materials, and every effort was made to replicate the technology that could be used in this decision situation. An in situ test over time would be the most desirable way to conduct the experimentation but would be very difficult to do. Finding 32 experienced subjects to participate voluntarily in the experiment was not an easy task, although the number represents a rather small sample size. The experiment was designed to test the performance of the subjects and not of the overall scores per model. If one assumes independence of each decision made by each subject, than the results would be even more significant.

Finding an optimal route between an origin and a destination on a transportation network can be viewed as a multiattribute decision problem where the alternatives (routes) are represented implicitly. The cognitive process of finding an optimal route can be based on two different approaches. One approach is to construct heuristically a small set of alternatives (routes), assess them, and choose the best one (decision making by alternative). The other approach is to assess each link with respect to the attributes, compute the value of all possible alternatives, and then choose the best one (decision making by attribute). Decision making by alternative in the case of a task such as routing vehicles on a network could be supported by an advanced user interface that facilitates the generation of alternatives with possible additions of a few simple heuristics. The results of this experiment suggest, however, that decision making by attribute can be more accurate and requires less effort. In addition, the decision support model must take into account the cognitive limitations of the user that affect human information processing in many operational settings.

The conclusion is that advanced information technology can improve operational risk management only by integrating appropriate decision models into its decision support systems. In addition, the results of this research encourage extending operational decision support to other fields of operational decision making. Whenever a sudden onset event occurs that can affect the attributes of scheduled and ongoing activities, planned courses of action must be reevaluated.

Since each subject worked with only one of the four reasoning logics, no conclusions can be drawn about whether a combination of the logics could yield better results. In fact, as Hulland and Kleinmuntz [1994] note, time pressure does not necessarily force one to employ the current choice strategy more rapidly, but it can cause a shift in choice strategy. For example, if events (such as snow storms) have been perceived to be risky, the conservative heuristic logic would be appropriate; but in instances where dangerous and harmless events are present simultaneously, assessment by attribute seems to be the most appropriate strategy. For non-complex rerouting situations, the dispatcher might select the new routes without the help of the decision support system.

Further research should focus on relaxing some of the stringent constraints that were used in this experimental setting. For example, having more operations (vehicles) and more RTEs would significantly increase the complexity of the tasks. This would make the scenarios more realistic but also include more parameters, requiring a more complex experiment to generate the necessary data.

Operational risk management has become technologically viable and headquarters are implementing advanced technologies to improve efficiency, safety, and security of their operations. However, the decision to change planned courses of action (such as emergency plans or transportation routes for hazardous materials) is not made solely by the operational risk manager, not only because other actors also have a say but also because they have indispensable expertise. The assessment, reassessment, and resulting suggestions for decision making must be made under pressure and time constraints.

Section 2 assessed cardinal and ordinal preference aggregation and decision making approaches for multiple operational risk managers (experts). The models are made adaptive by using dynamically changing weights and consistency coefficients over time, and the experts' aggregated assessments are computed iteratively. At each iteration step, some of the experts are offered the possibility to reconsider their assessment. However, there might not be time for this quasi real-time Delfi, approach and the models have to progress without feedback, by eliminating step-by-step outliers. Weights of the experts are adjusted continuously but could also be pre-set at any time.

The application of these models to field data shows their realism. Both assessment and rerouting decisions complied with what a majority of the individual experts suggested. However, the real value of a multi-expert operational aggregation procedure must be assessed in a quasi field setting, where the unavailability of some experts during some RTEs, the experts' reluctance to make or revise assessments, and the different reactions to time delay and lack of information and data will indicate future directions for research. The effects of feedback and learning must also be investigated. Feedback accounts for a relative improvement of the group assessment by showing the experts the assessments and choices of the other experts. Learning, on the other hand, accounts for an absolute improvement by instructing the experts about the consequences of their assessments and decisions.

Despite the fact that more investigations must extend this work before firm conclusions can be drawn, the results of this research can have an impact on the design of human-machine systems and on the tasks of organizations focusing on centralized real time decision support for remote and mobile units. Efforts in this directions are going on in different domains all over the world. The results of this research show that the appropriate reasoning approach, embedded into a decision support system, can reduce effort and improve accuracy in real time decision making.

Thus, the steadily growing use of real-time decision making technology by industry (generally regardless of the economic benefits), coupled with the fact that decision makers tend to adapt their decision making strategy to the decision aid that reduces effort the most (generally regardless of accuracy) [Todd and Benbesat, 1991], motivates further investigations of real time decision support.

Advanced communications and human-machine interface technologies, especially mobile communications and the Internet, are being developed that will enable us to "see," "hear" and "feel" the operations of decentralized stationary and mobile systems and their surroundings. However, there still will be the need for ways of processing these data into information in order to make us "smarter." Models that provide the appropriate cognitive support offer one way to do this, and they are the raison d'être for the research reported in this volume.

REFERENCES

A) References to Authors' Work (by Chapters)

Chapter I
Beroggi G.E.G., 1997. "Satellite Communications and Expert Systems." In *Advances in Expert Systems and Artificial Intelligence for Management*: Grabowski M. and Wallace W.A. (eds.), JAI Press.

Beroggi G.E.G. and Wallace W.A., 1994. "Operational Risk Management: A New Paradigm for Risk Analysis." *IEEE Transactions on Systems, Man, and Cybernetics*, 24/10, 1450-1457.

Beroggi G.E.G. and Wallace W.A. 1994. "Global Logistics Using Satellite Technology for Monitoring and Control." In *Global Information Systems and Technology: Focus on the Organization and Its Functional Areas*. Deans P.C. and Karwan K.R. (eds.), Idea Group Publishing, Harrisburg, Pa., 329-341.

Beroggi G.E.G. 1993. "Employing Advanced Communications and Computing Technology to Benefit from Changing Freight Transportation Policies." In *Advanced Technologies*, Beheshti M.R. and Zreik K. (eds.), Elsevier, New York, 341-348.

Beroggi G.E.G. and Wallace W.A., 1992. "Real-Time Control of the Transportation of Hazardous Materials." *Journal of the Urban and Regional Information Systems Association*, 4/1, 56-65.

Beroggi G.E.G. and Wallace W.A., 1991. "Closing the Gap - Transit Control for Hazardous Material Flow." *Journal of Hazardous Materials*, 27, 61-75.

Chapter II
Beroggi G.G.E., 1994. "A Real-Time Routing Model for Hazardous Materials." *European Journal of Operational Research*, 75/3, 508-520.

Beroggi G.E.G. and Wallace W.A., 1994. "Operational Risk Management: A New Paradigm for Risk Analysis." *IEEE Transactions on Systems, Man, and Cybernetics*, 24/10, 1450-1457.

Beroggi G.E.G. and Wallace W.A. 1994. "A Decision Logic for Operational Risk Management. In *Computational Organization Theory*. Carley K.M. and Prietula M.J. (eds), Hillsdale, NJ, Lawrence Elbaum Associate, 289-308.

Chapter III
Beroggi G.E.G., 1997. "Satellite Communications and Expert Systems." In *Advances in Expert Systems and Artificial Intelligence for Management*: Grabowski M. and Wallace W.A. (eds.), JAI Press.

Beroggi G.E.G., Aebi M., and Wallace W.A., 1995. MEMIS: Multimedia Emergency Management Information System." Globalization of Emergency Management and Engineering: National and International Issues Concerning Research and Applications. Sullivan J.D., Wybo J.-L., and Buisson L. (eds.), The International Emergency Management and Engineering Conference, May 9-12, Nice (F), 443-448.

Beroggi G.E.G, Waisel L., and Wallace W.A., 1995. "Employing Virtual Reality Technology to Support Decision Making in Emergency Management." *Safety Science*, 20, 79-88.

Beroggi G.E.G. and Wallace W.A. (eds.), 1995. *Computer Supported Risk Management*. Kluwer Academic Publishers, Amsteradam, ISBN: 0-7923-3372-1, 372 pages.

Wallace W.A., 1992. "Distributed Decision Making and Group Collaboration over Networks using Virtual Vnvironments." Technical Report, Decision Sciences and Engineering Systems, Rensselaer Polytechnic Institute, Troy, New York.

Chapter IV

Beroggi G.E.G., 1995. "On the Move: Monitoring and Routing Freight with Satellite Systems and GIS." *GIS-Europe*, 4/1, 24-26.

Beroggi G.E.G. and Wallace W.A. 1994. "A Prototype Decision Support System in Hypermedia for Operational Control of Hazardous Material Shipments." *Decision Support Systems*, 12, 1-12.

Beroggi G.G.E., 1994. "A Real-Time Routing Model for Hazardous Materials." *European Journal of Operational Research*, 75/3, 508-520.

Beroggi G.E.G., Hersperger A., Wallace W.A., Wiedmer M. und Zumsteg M.B., 1993. *Operationelle Routenwahl im Gefahrenguttransport: Methodische Ansätze, Computerimplementation, experimentelle Bewertung*. Verlag der Fachvereine, Zürich, ISBN: 3-7281-2061-8, 98 pages.

Beroggi G.E.G. and Wallace W.A., 1992. "Real-Time Control of the Transportation of Hazardous Materials." *Journal of the Urban and Regional Information Systems Association*, 4/1, 56-65.

Beroggi G.E.G., 1991. *Modelling Real-Time Decision Making for Hazardous Material Transportation*. Ph.D. Dissertation, Rensselaer Polytechnic Institute, Troy, New York. University Microfilm International, No. 9202173, 300 N Zeeb Road, Ann Arbor, MI (275 pages).

Chapter V

Beroggi G.E.G. and Wallace W.A., 1995. "Real-Time Decision Support for Emergency Management." *Journal of Contingengies and Crises Management*, 3/1, 18-26.

Beroggi G.E.G. and Wallace W.A., 1994. "Operational Risk Management: A New Paradigm for Risk Analysis." *IEEE Transactions on Systems, Man, and Cybernetics*, 24/10, 1450-1457.

Chapter VI

Ikeda Y., Beroggi G.E.G., and Wallace W.A., 1995. "Tactical Air Raiding: An Application of Operational Risk Management." Proceedings of the Simulation MultiConference, April 9-13, Phoenix, Arizona, Military, Government, and Aerospace Simulation, Chinni M.J. (ed.), Simulation Series, 27/4, Society for Computer Simulation, ISBN: 1-56555-075-7.

Ikeda Y., Beroggi G.E.G., and Wallace W.A., (forthcoming). "Supporting Multi-Group Emergency Management with Multimedia." *Safety Science*.

Ikeda Y., Beroggi G.E.G., and Wallace W.A., 1997. "Multimedia Support for Multi-Group Operational Risk Management." In Hevner A.R. and Koehler G.J. (eds.) Proceedings of the Second INFORMS Conference on Information Systems and Technology, 174-181.

Chapter VII

Beroggi G.E.G. and Wallace W.A., 1995. "Operational Control of the Transportation of Hazardous Materials: An Assessment of Alternative Decision Models." *Management Science*, 41/12, 1962-1977.

Beroggi G.E.G. and Wallace W.A., 1997. "Decision Support for Hazardous Operations: The Effect of Reasoning Logics on Decision Making." *IEEE Transactions on Systems, Man, and Cybernetics*.

Beroggi G.E.G., Hersperger A., Wallace W.A., Wiedmer M. und Zumsteg M.B., 1993. *Operationelle Routenwahl im Gefahrenguttransport: Methodische Ansätze, Computerimplementation, experimentelle Bewertung*. Verlag der Fachvereine, Zürich, ISBN: 3-7281-2061-8, 98 pages.

B) References in Text

AAA, 1986; Foundation for Traffic Safety. "Local Response to Hazardous Materials Incidents and Accidents." *Transportation Quarterly*, 40/4, 461-482.

Abkowitz M and Cheng P.D.-M., 1988. "Developing a Risk/Cost Framework for Routing Truck Movement of Hazardous Materials." *Accident Analysis and Prevention*, 20, 39-51.

Apostolakis G., 1978. "Probability and Risk Assessment: The Subjective Viewpoint and some Suggestions." *Nuclear Safety*, 19/3, 305-315.

Arrow K.J., 1951. *Social Choice and Individual Values*. 1st edition, 2nd edition 1963. John Wiley, New York.

BAP, 1988; Schweizerisches Bundesamt für Polizeiwesen. Leitfaden für den Transport gefährlicher Güter auf der Strasse, Eidgenössische Drucksachen- und Materialzentrale, Bern, Switzerland.

Batz T.M., 1991. "The Utilitzation of Real-Time Traffic Information by the Trucking Industry." *IEEE Transactions on Vehicular Technology*, 40, 64-67.

Beaton R.M., Addams M.B., and Harrison J.V.A., 1987. "Real-Time Mission and Trajectory Planning," Proceedings of the 26th Conference on Decision and Control," IEEE, Piscataway, NJ, 1954-1959.

Belardo, S., Karwan K.R., and Wallace W.A., 1984. "Managing the Response to Disasters Using Microcomputers," *Interfaces*, 14/2, 29-39.

Ben Zur H. and Breznitz J., 1981. "The Effect of Time Pressure on Risky Choice Behavior." *Acta Psychologica*, 47, 89-104.

Behar R., 1989. "Joe's Bad Trip." *TIME Magazine*, July 24, 42-47.

Beinat E., Nijkamp P., Rietveld P., 1994. "Value Functions for Environmental Pollutants: A Technique for Enhancing the Assessment of Expert Judgements." *Environmental Monitoring and Assessment*, 30, 9-23.

Boffey T.B. and Karkazis J., 1995. "Linear versus Nonlinear Models for Hazmat Routing." *INFOR*, 33/2, 114-117.

Boghani A.B. and Mudan K.S., 1987. "Comparative Risk Assessment of Transportation of Hazardous Materials in Urban Areas." Arthur D. Little, Inc. Cambridge, Massachusetts. Presented at the Society for Risk Analysis 1987 Annual Meeting Houston, Texas.

Bond G., 1988. *XCMD's for HyperCard*. MIS Press, Portland, Oregon.

Borsook P., 1994. "Data Communications." *IEEE Spectrum*, January, 26-29.

Briskman R.D., 1988. "Vehicle Location by Satellite." International Congress on Transportation Electronics, 247-249.

Brodley R.F., 1982. "The Combination of Forecasts: A Bayesian Approach." *Journal of the Operational Research Society*, 33, 171-174.

Brown G.G. and Graves G.W., 1981. "Real-Time Dispatch of Petroleum Tank Trucks." *Management Science*, 27/1, 19-32.

Brugioni D.A., 1988. "Aerial Reconnaissance and Photo Interpretation in Emergency Management." Symposium on Information Technology and Emergency Management, Tennessee, 122-126.

Buchmann, P., 1988. "Vehicle Communication in Europe." International Congress on Transportation Electronics, 229-235.

Burger J., 1995. *Multimedia for Decision Makers - A Business Primer*. Addison Wesley.

Burns R.D, 1989. "Safety and Productivity Improvement of Railroad Operations by Advanced Train Control Systems." IEEE/ASME Joint Conference, 33-38.

BUWAL, 1991. Handbuch zur Störfallverordnung. Bundesamt für Umwelt, Land und Landschaft, Bern, Switzerland.

CAMEO, 1993. Computer Aided Management of Emergency Operations. National Safety Council, Environmental Health Center, 1019 19th Street, N.W., Suite 401, Washington, D.C. 20036.

CCPS, 1992. *Guidelines for Hazard Evaluation Procedures*. 2nd edition, Center for Chemical Process Safety of the American Institute of Chemical Engineers, New York.

Center for Chemical Process Safety, Guidelines for Hazard Evaluation Procedures. American Institute of Chemical Engineers, New York, New York, 1992, p. 9.

Chidambaram L., Jones B., 1993. "Impact of Communication Medium and Computer Support on Group Peceptions and Performance: A Comparison of Face-to-Face and Dispersed Meetings." *MIS Quarterly,* 17, 465-488.

Clemen R.T., 1989. "Combining Forecasts: A Review and Annotated Bibliography." *International Journal of Forecasting*, 5, 559-583.

COM(90)490, 1990. Towards Europe-Wide Systems and Services - Green Paper on a Common Approach in the Field of Satellite Communications in the European Community. Communications from the Commission of the European Communities.

Conover W.J., 1980. *Practical Nonparametric Statistics*. John Wiley & Sons, New York.

Cooke R.M., 1991. *Experts in Uncertainty: Opinion and Subjective Probability in Science*. Oxford University Press, Oxford.

Coulouris G. and Thimbleby H., 1993. *HyperProgramming: Building Interactive Programs with HyperCard*. Addison-Wesley, New York.

Cuena J., 1988. "Building Expert Systems Based on Simulation Models: An Essay in Methodology." Facultad de Informatica, Universidad Politecnica Madrid, Spain.

Dantzig G.B., 1975. "On the Shortest Route Through a Network." *The Mathematical Association of America, Studies in Graph Theory*, Part I, 89-93.

DeKeyser V., 1987. "How can Computer-Based Visual Displays Aid Operators?" *International Journal of Man-Machine Studies*. 27, 471-478.

Dess G.G., Rasheed A.M.A., McLaughlin K.J., and Priem R.L., 1996. "The New Corporate Architecture." *IEEE Engineering Management Review*, 24/2, 20-28.

DeWispelare A.R., Herren L.T., and Clemen R.T., 1995. "The Use of Probability Elicitation in the High-Level Nuclear Waste Regulation Program." *International Journal of Forecasting*, 11, 5-24.

Einhorn H.J. and Hogarth R.M., 1981. "Behavioral Decision Theory: Process of Judgment and Choice." *Annual Review of Psychology*, 32, 53-88.

Egsegian R., Pittman K., Farmer K., and Zobel R., 1993. "Practical Applications of Virtual Reality to Firefighter Training." In: Sullivan, J.D. (Ed.), Proceedings of the 1993 Simulation Multiconference on the International Emergency Management and Engineering Conference, April, 155-160.

ESA, 1993. Europe at the Crossroads: The Future of Its Satellite Communications Industry. Final Report of the Working Group on Satellite Communications Policy, European Space Agency, ESA SP-1166.

Fischhoff B., Lichtenstein S., Slovic P., Derby, and Keeney R.L., 1981. *Acceptable Risk*. Cambridge University Press.

Gibbons A., 1989. *Algorithmic Graph Theory*. Cambridge University Press.

Glickman T.S. and Sontag M.A., 1995. "The Tradeoffs Associated with Rerouting Highway Shipments of Hazardous Materials to Minimize Risks." *Risk Analysis*, 15/1, 61-67.

Glickman T.S., 1991. "An Expeditious Risk Assessment of the Highway Transportation of Flammable Liquids in Bulk." *Transportation Science*, 25/2, 115-123.

Glickman T.S., 1989. "Flammable Liquid Transportation Risks: A Case Study of Tank Trucks on Urban Roads." Resources for the Future, Washington, D.C.

Glickman T.S., 1988. "Hazardous Materials Routing - Risk Management or Mismanagement?" Resources, No. 3, Fall 1988, pp 11-13.

Grenier J., 1996. "Eutelsat: Towards the 21st Century." In Turner A. (ed.), *The International Telecommunications Update 1996/97*. Kensington Publications Ltd., London, 293-297.

Grierson N., 1996. "The Evolution of the PC as a Communications Device." In Turner A. (ed.), *The International Telecommunications Update 1996/97*. Kensington Publications Ltd., London, 359-361.

Haddow G.D., 1987. "The Safe Transportation of Hazardous Materials." *Transportation Quarterly*, 41/3, 381-322.

Harrald, J.R., Cohn, R., and Wallace, W.A., 1992. "We Were Always Re-Organizing...: Some Crisis Implications of the Exxon Valdez Oil Spill." *Industrial Crisis Quarterly*, 6, 197-217.

Harrald J.R., Marcus H., and Wallace W.A., 1990. "The Exxon Valdez: An Assessment of Crisis Prevention and Management Systems," *Interfaces* 25/5, 14-30.

Harris C.B., Goss R., Krakiwsky J., and Karimi H.A., 1987. "Optimal Route Information System - Automatic Vehicle Location (AVL) Route Determination." Department of Surveying Engineering, University of Calgary, Canada.

Hatcher M., 1995. "Introduction to Multimedia Supported Group/Organizational Decision System." *Decision Support Systems*, 15, 179-180.

Hillier F.S. and Lieberman G.J., 1986. *Introduction to Operations Research*. Holden0Day, Inc., fourth edition, Oakland, CA.

Hinz W., Arnoldt C., and Hessler C., 1993. "Screen-Based Process Control in Nuclear Power Plants." *Kerntechnik*, 58/2, 98-103.

Hopkin V., 1989. "Man-Machine Interface Problems in Designing Air Traffic Control Systems," Proceedings of the IEEE, Special Issue on Air Taffic Control, November.

Hulland J.S. and Kleinmuntz D.N., 1994. "Factors Influencing the Use of Internal Summary Evaluations Versus External Information in Choice." *Journal of Behavioral Decision Making*, 7, 79-102.

InterClair, 1992. Environmental Models for Risk Assessment and Management of Air and Surface Water Pollution in Large Industrial Complexes. International Atomic Energy Agency (IAEA), Vienna.

ITS, 1995. The Second International World Congress on Intelligent Transport Systems, Yokohama, November 9-11, Volume IV, T15.3, Routing/Scheduling, 2027-2061.

Janis I. L. and Mann L., 1977. *A Psychological Analysis of Conflict, Choice, and Commitment*. Free Press, New York.

Johnson E.J. and Payne J.W., 1985. "Effect and Accuracy in Choice." *Management Science*, 31, 395-414.

Jurgen R.K., 1991, "Smart Cars and Highways go Global." *IEEE Spectrum*, Special Report, May, 26-36.

Kalelkar A.S. and Brooks R.E., 1978. "Use of Multidimensional Utility Functions in Hazardous Shipment Decisions." *Accident Analysis and Prevention*, 10, 251-265.

Karimi H.A., and Krakiwsky E.J., 1988. "Design Concepts for Knowledge-Based Route Guidance Systems," Proceedings: Position, Location, and Navigation Symposium, IEEE, Piscataway, NJ, 95-103.

Keen K., 1992. "European Community Research and Technology Development on Advanced Road Transport Telematics, 1992-1994." *Traffic Engineering and Control*, April, 263-267.

Kendall M. and Gibbons J.D., 1990. *Rank Correlation Methods*. Fifth Edition, Oxford University Press, New York.

Keyes J., 1996. "Intelligent Agent Technology: The Mistery Explained." In Turner A. (ed.), *The International Telecommunications Update 1996/97*. Kensington Publications Ltd., London, 244-246.

Kleindorfer P.R., Kunreuther H.C., and Schoemaker J.H., 1993. Decison Sciences: An Integrated Perspective. Cambridge University Press, New York.

Kraay D., Harker P., and Chen B., 1991. "Optimal Pacing of Trains in Freight Railroads: Model Formulation and Solution." *Operations Research*, 39/1, 82-99.

Lange L., 1998. "The Internet." *IEEE/Spectrum*, January, 37-42.

List G.F., Mirchandani P.B., Turnquist M.A., and Zografos K.G., 1991. "Modeling and Analysis for Hazardous Materials Transportation: Risk Analysis, Routing/Scheduling, and Facility Location." *Transportation Science*, Special Issue, 25/2, 100-114.

Mancini G., 1987. "Commentary: Models of the Decision Maker in Unforseen Accidents." *International Journal of Man-Machine Studies*, 27, 631-639.

Marchioni G. and B. Schneiderman, 1988. "Finding Facts vs. Browsing Knowledge in Hypertext Systems." *IEEE Computer*, January, 70-80.

McCord M.R. and Leu Y.-C., 1995. "Sensitivity of Optimal Hazmat Routes to Limited Preference Specification." *INFOR*, 33/2, 68-83.

Mitchell J.K., 1990. "Human Dimensions of Environmental Hazards: Complexity, Disparity and the Search for Guidance." In Kirby A. (ed.), *Nothing to Fear*, University of Arizona Press.

Moore R.L., Hawkinson L.B., Levin M., Hofmann A.G., Matthews R.L., and David M.H., 1988. "Expert Systems Methodology for Real-Time Process Control," Gensym Corporation, Cambridge, MA.

Moray N., 1987. "Intelligent Aids, Mental Models, and the Theory of Machines." *International Journal of Man-Machine Studies*, 27, 619-629.

Morentz J.M., 1989. "An Automated Crisis Response System for Hazardous Materials Transportation Emergency in New Jersey." Division of Environmental Quality Department of Environmental Protection, state of New Jersey.

Morlok E.K. and Halowell S.F., 1989. Reported Benefits of Advanced Vehicle Tracking and Communications Systems. Mobile Satellite Project, WP-89-8-1, Department of Systems, University of Pennsylvania, Philadelphia.

Myung I.J, Ramamoorti S., Bailey A.D. Jr., 1996. "Maximum Entropy Aggregation of Expert Predictions." *Management Science*, 42/10, 1420-1436.

Naum S.R., Roy A., and Kumara S., 1989. "A Decision Support System for Real-Time Control and Monitoring of Dynamic Processes," Proceedings American Control Conference, IEEE, Piscataway, NJ, 361-365.

OTA, 1986. "Transportation of Hazardous Materials: State and Local Activities." Office of Technology Assessment, Washington, D.C.

Ostroff J.S. , 1989. *Temporal Logic for Real-Time Systems*. Somerset, UK: Research Studies Press Ltd.

Payne E.C. and R.C. McArthur, 1990. *Developing Expert Systems: A Knowledge Engineer's Handbook for Rules & Objects*. John Wiley & Sons, New York.

Payne J.W., Bettman J.R., and Johnson E.J., 1993. The Adaptive Decision Maker. Cambridge University Press, New York.

Payne J.W., Bettman J.R., and Johnson E.J., 1988. "Adaptive Strategy Selection in Decision Making." *Journal of Experimental Psychology: Learning, Memory, and Cognition*, 14, 534-552.

Payne J.W., 1987. "Contingent Decision Behavior," *Psychological Bulletin*, 92, 382-402.

Payne J.W. and Braunstein M.L., 1978. "Risky Choice: An Examination of Information Acquisition Behavior." *Memory & Cognition*, 6, 554-561.

Payne J.W., 1976. "Task Complexity and Contingent Processing in Decision Making: An Information Search and Protocol Analysis." *Organizational Behavior and Human Performance*, 16, 366-387.

Peckham R.J., Haastrup P., and Otway H., 1988. "A Computer-Based System for Risk Management Support." *Decision Support Systems*, 4, pp 481-489.

Perrow C., 1989. *Normal Accidents - Living with High-Risk Technologies*. Basic Books, Inc., Publishers, New York.

Phillips D., 1998. "Transportation." *IEEE/Spectrum*, January, 84-89.

Powell W.B., 1990. "Real-Time Optimization for Truckload Motor Carriers," *OR/MS Today*, April, 28-32.

Pritchard W.L., Suyderhoud H.G., and Nelson R.A., 1993. *Satellite Communication Systems Engineering*. PTR Prentice Hall, Englewood Cliffs, New Jersey.

Psotka J., and Davison S., 1993. Virtual Reality Terms. Electronic Communication.

Ramanathan S., Rangan P.V., and Harrick M.V., 1992. Integrating Virtual Reality, Teleconferencing, and Entertainment into Multimedia Home Computers." *IEEE Transactions on Consumer Electronics*, 38/2, 70-76.

Ramesh B., Sengupta K., 1995. "Multimedia In A Design Rationale Decision Support System." *Decision Support Systems,* 15, 181-196.

Rheingold W., 1993. *Virtual Reality.* Cox & Wyman Ltd, Reading , Berks, UK.

Rheingold H., 1991. *Virtual Reality.* Touchstone Press, New York.

Riezenman M.J., 1998. "Communications." *IEEE/Spectrum*, January, 29-36.

Robinett W., 1992. *Synthetic Experience. Presence*, 1/2, 229-247.

Roche E.M., 1995. "Business Value of Electronic Commerce Over Interoperable Networks." *Information Infrastructure and Policy*, 4, 307-325.

Russell E.R., Smaltz J.J., Lambert J.P., Deines V.P., Jepsen R.L., Joshi P.G., and Mansfield T.R., 1986. "Risk Assessment Users Manual for Small Communities and Rural Areas." U.S. Department of Transportation, University Research Program, Washington, D.C.

Saccomanno F.F. and Chan A.Y.-W., 1985. "Economic Evaluation of Routing Strategies for Hazardous Road Shipments." *Transportation Research Record*, 1020, 12-18.

Sage A.P., 1986. "Information Technology for Crisis Management." *Large Scale Systems*, 11, 193-205.

Sandri S.A., Dubois D., and Kalfsbeek H.W., 1995. "Elicitation, Assessment, and Pooling of Expert Judgments Using Possibility Theory." *IEEE Transactions on Fuzzy Systems*, 3/3, 313-335.

Schneiderman R., 1994. *Wireless Personal Communications: The Future of Talk.* IEEE Press, Inc., New York.

Schrijver P.R. and Sol H.G., 1992. "Planning Support for Real-Time Management of Road Transportation." In *Decision Support Systems: Experiences and Expectations.* T. Jelassi, M.R. Klein and W.M. Mayon-White, eds., Elsevier Science Publishers B.V. (North-Holland), 115-132.

Schutte P.C., Abbott K.H., Palmer M.T., and Ricks W.R., 1987. "An Evaluation of a Real-Time Fault Diagnosis Expert System for Aircraft Applications," Proceedings of the 26th Conference on Decision and Control, IEEE, Piscataway, NJ, 1941-1947.

Sheffi Y., 1991. "A Shipment Information Centre," *The Journal of Logistics Management*, 2,/2, 1-12.

Sheridan T.B., 1992. Defining Our Terms. *Presence*, 1/2, 272-273.

Simon H.A., 1972. "Theories of Bounded Rationality." In McGuire C.B. and Radner R. (eds.), *Decision and Organization.* University of Minnesota Press, Minneapolis, 161-176.

Smart C. and Vertinsky J., 1980. "Designs for Crisis Decision Units." *Administrative Science Quarterly*, 22/4, 640-657.

Smith P.A. and Wilson J.R., 1993. "Navigation in Hypertext Through Virtual Environments." *Applied Ergonomics*, 24/4, 271-278.

Tanimoto S.L., 1987. *The Elements of Artificial Intelligence*. Computer Science Press, Inc., Rockville, Maryland.

Tirró S., 1993. *Satellite Communication Systems Design*. Plenum Press, New York.

Todd P. and Benbesat I., 1991. "An Experimental Investigation of the Impact of Computer Based Decision Aids on Decision Making Strategies." *Information Systems Research*, 2, 87-115.

Turnquist M.A. and Zografos, K.G., 1991 (eds.). "Transportation of Hazardous Materials." *Transportation Science, 1991*, Special Issue, 25/2.

Urbanek G.L. and Barber E.J., 1980. "Development of Criteria to Designate Routes for Transporting Hazardous Materials." Federal Highway Administration, U.S. Department of Transportation, Washington, D.C.

U.S. GAO, 1980. "Programs for Ensuring the Safe Transportation of Hazardous Materials Need Improvement." General Accounting Office, Reported by the Comptroller General of the United States.

Wagenaar W. and Groeneweg J, 1987. "Accidents at Sea: Multiple Causes and Impossible Consequences." *International Journal of Man-Machine Studies*. 27, 587-598.

Wells M., 1992. Virtual Reality: Technology, Experience, Assumptions. *Human Factors Society Bulletin*, 35/9, 1-3.

West N., 1993. "Multimedia Masters: A Guide to the Pros and Cons of Seven Powerful Authoring Programs." *Macworld*, March, 114-117.

Witteman P.A., 1989. "The Stain Will Remain on Alaska." *TIME Magazine*, September 25, 1989.

Yager R.R., 1980. "On a General Class of Fuzzy Connectives." *Fuzzy Sets and Systems*, 4, 235-242.

Yankelovich N., Haan B.J., Meyrowitz N.K., and Drucker S., 1988. "Intermedia: The Concept and the Construction of a Seamless Information Environment," *IEEE Computer*, January, 81-96.

Zakay D., 1985. "Post-Decisional Confidence and Conflict Experience in a Choice Process." *Acta Psychological*, 58, 75-80.

Zeltzer, D., 1992. "Autonomy, Interaction, and Presence." *Presence*, 1/2, 127-132.

Subject Index

activity
    affected, 54
    matrix, 42
    pre-event, 1
    post-event, 1
    relevant, 54
air raid command, 140
assessment
    acceptability, 67
    conservative heuristic, 165
    efficient, 67
    experimental, 163
    multiattribute utility, 166
    ordinal preference, 165
    proper, 66
    satisfactory, 67
    strategic, 105
    sufficient, 66
    real-time, 195
    visual interactive, 165
attributes, 43
AWACS, 141
axiom, 52

clause, 57
cognitive
    assumptions, 47
    load, 183
communications
    satellite, 12
    systems, 12
    technology, 15
connected, 41
consistency
    group, 64
    long-run, 63
    measure, 63
control
    operational, 8
    transit, 9
    variables of, 9
course of action, 38
    analogous, 68
    congruent, 68
    connected, 41
    feasible, 41
    finding, 53
    indifferent, 68
    optimal, 132
    revising, 53, 131
    sub-, 40
cyberspace, 92

decision
    maker
        individual, 47
        multiple, 60
    making, 105, 137
        by alternative, 165
        by attribute, 165
        individual, 47
        multiple groups, 155
        time frame, 105
    logic
        air raid command, 142
        emergency response, 128
        energy generation, 154
        HazMat transport, 107
        reasoning, 51
    support system, 112
deviation
    coefficient, 63
    relative, 63
disjoint
    activity-, 41
    decision-, 41
dispatcher, 101

Eutelsat, 16
evaluation by
    alternative, 48
    attribute, 48
experiment, 169

generic tasks, 3, 9, 29
Global Positioning System, 16

human-machine, 10, 36
history,
    operational, 50
    state, 50
hypothesis
    accuracy, 169, 182
    effort, 168

identification, 102
incident
    command center, 125
    from, 40
    to, 40
Inmarsat, 15
Internet, 78
Iridium, 13, 21

Kendall's coefficient of
    concordance, 64
knowledge-based, 13

LP-HC, 1

media
    hyper-, 71
    multi-, 71, 171
monitoring, 102
multiattribute utility model, 57

nuclear power, 151

oil spill, 133
operational risk management
    active, 53
    concept, 24
    emergency response, 124
    environment
        air raid command, 140
        emergency response, 124
        energy generation, 152
        HazMat transport, 100
    HazMat, 115
    practicality, 31
    mathematics of, 38
    motivation, 3
    passive, 53

positioning, 12
possibility, 31
predicate, 57
    logic, 57
preference
    algebra, 51
    assessment, 58, 137
    cardinal, 62
    conditional, 56
    classes, 46
        model
            multiattribute utility, 108
            ordinal, 108
        graph, 47
        ordinal, 66
        unconditional, 55
prestructured, 41
probability, 31

range of acceptability, 64
RARC, 140
real-time event (RTE), 38
    area, 103
    point, 103

reasoning, 9
    logic, 51
recovery, 1
result
    accuracy, 178, 181
    effort, 177, 181
risk
    management
        operational vs. strategic, 31
route
    robustness, 122
    selection process, 118
routing
    algorithm, 11, 107, 108, 111, 118
    re-, 103, 107, 111, 114, 118, 122

SAM, 141
satellite
scenario, 121, 170
sensing, 9
sensitive, 42
solvability, 42
    index, 43
stable, 42
subject, 173

task
    complexity, 182
    in experiment, 170
tele
    operation, 98
    presence, 98
time
    contextual aspect, 51
    max for assessment, 106, 116
    min for assessment, 106, 116
    spatial aspect, 50
topological graph, 39
    emergency response, 129
transportation
    modes, 11
    of hazardous materials, 100

vehicles
    affected
        management, 106
        rerouting, 103
videoconferencing, 85
virtual
    instrument, 93
    reality, 89

weight, 62